Stop Faking It!

Finally Understanding Science So You Can Teach It

ENERGY

Stop Faking It!

Finally Understanding Science So You Can Teach It

ENERGY

By William C. Robertson, PhD

National Science Teachers Association
Arlington, Virginia

Claire Reinburg, Director
Judy Cusick, Associate Editor
Carol Duval, Associate Editor
Betty Smith, Associate Editor

ART AND DESIGN Linda Olliver, Director
Brian Diskin, Illustrator
PRINTING AND PRODUCTION Catherine Lorrain, Director
Nguyet Tran, Assistant Production Manager
Jack Parker, Desktop Publishing Specialist
PUBLICATIONS OPERATIONS Erin Miller, Manager

MARKETING Holly Hemphill, Director
NSTA WEB Tim Weber, Webmaster
PERIODICALS PUBLISHING Shelley Carey, Director
SciLINKS Tyson Brown, Manager

NATIONAL SCIENCE TEACHERS ASSOCIATION
Gerald F. Wheeler, Executive Director
David Beacom, Publisher

1840 Wilson Blvd., Arlington, VA 22201
www.nsta.org/store
For customer service inquiries, please call 800-277-5300.

Energy: Stop Faking It! *Finally Understanding Science So You Can Teach It*
NSTA Stock Number: PB169X2
18 17 16 10 9 8

Library of Congress Cataloging-in-Publication Data
Robertson, William C.
Energy / William C. Robertson.
p. cm. — (Stop faking it! : finally understanding science so you can teach it)
ISBN 0-87355-214-8
1. Force and energy. 2. Power (Mechanics) I. Title.
QC73 .R55 2002
531'.6—dc21 2002010454

Featuring sciLINKS®—a way of connecting text and the Internet. Up-to-the-minute online content, classroom ideas, and other materials are just a click away. Go to page x to learn more about this educational resource.

eISBN 978-1-93353-184-7

Contents

How can you and your students avoid searching hundreds of science web sites to locate the best sources of information on a given topic? SciLinks, created and maintained by the National Science Teachers Association (NSTA), has the answer.

In a SciLinked text, such as this one, you'll find a logo and keyword near a concept your class is studying, a URL (*www.scilinks.org*), and a keyword code. Simply go to the SciLinks web site, type in the code, and receive an annotated listing of as many as 15 web pages—all of which have gone through an extensive review process conducted by a team of science educators. SciLinks is your best source of pertinent, trustworthy Internet links on subjects from astronomy to zoology.

Need more information? Take a tour—*http://www.scilinks.org/tour*

Preface

When I was back in college, there was a course titled Physics for Poets. At a school where I taught physics, the same kind of course was referred to by the students as Football Physics. The theory behind having courses like these was that poets and/or football players, or basically anyone who wasn't a science geek, needed some kind of watered-down course because most of the people taking the course were—and this was generally true—SCARED TO DEATH OF SCIENCE.

In many years of working in education, I have found that the vast majority of elementary school teachers, parents who home school their kids, and parents who just want to help their kids with science homework fall into this category. Lots of "education experts" tell teachers they can solve this problem by just asking the right questions and having the kids investigate science ideas on their own. These experts say you don't need to understand the science concepts. In other words, they're telling you to fake it! Well, faking it doesn't work when it comes to teaching *anything,* so why should it work with science? Like it or not, you have to understand a subject before you can help kids with it. Ever tried teaching someone a foreign language without knowing the language?

The whole point of the *Stop Faking It!* series of books is to help you understand basic science concepts and to put to rest the myth that you can't understand science because it's too hard. If you haven't tried other ways of learning science concepts, such as looking through a college textbook, or subscribing to *Scientific American,* or reading the incorrect and oversimplified science in an elementary school text, please feel free to do so and then pick up this book. If you find those other methods more enjoyable, then you really are a science geek and you ought to give this book to one of us normal folks. Just a joke, okay?

Just because this book series is intended for the non-science geek doesn't mean it's watered-down material. Everything in here is accurate, and I'll use math when it's necessary. I will stick to the basics, though. My intent is to provide a clear picture of underlying concepts, without all the detail on units, calculations, and intimidating formulas. You can find that stuff just about anywhere.

Also, I'll try to keep it lighthearted. Part of the problem with those textbooks (from elementary school through college) is that most of the authors and the teachers who use them take themselves way too seriously. I can't tell you the number of times I've written a science curriculum only to have colleagues tell me it's "too flip" or, "You know, Bill, I just don't think people will get this joke." Actually, I don't really care if you get the jokes either, as long as you manage to learn some science here.

Speaking of learning the science, I have one request as you go through this book. There are two sections titled *Things to do before you read the science stuff* and *The science stuff.* The request is that you actually DO all the "things to do" when I ask you to do them. Trust me, it'll make the science easier to understand, and it's not like I'll be asking you to go out and rent a superconducting particle accelerator. Things around the house should do the trick. If you are a classroom teacher, you might be tempted to do a number of the activities in this book with your students. If you do that, use a bit of common sense when it comes to safety. Ask yourself if you really want your students using open flames and the like!

By the way, the book isn't organized this way (activities followed by explanations followed by applications) just because it seemed a fun thing to do. This method for presenting science concepts is based on a considerable amount of research on how people learn best and is known as the *Learning Cycle*. There are actually a number of versions of the Learning Cycle but the main idea behind them all is that we understand concepts best when we can anchor them to our previous experiences.

One way to accomplish this is to provide the learner with a set of experiences and then explain relevant concepts in a way that ties the concepts to those experiences. Following that explanation with applications of the concepts helps to solidify the learner's understanding. The Learning Cycle is not the only way to teach and learn science, but it is effective in addition to being consistent with recommendations from *The National Science Education Standards* (National Research Council 1996) on how to use inquiry to teach science. (Check out Chapter 3 of the *Standards* for more on this.) In helping your children or students to understand science, or anything else for that matter, you would do well to use this same technique.

As you go through this book, you'll notice that just about everything is measured in *Système Internationale* (SI) units, such as meters, kilometers, and kilograms. You might be more familiar with the term *metric units,* which is basically the same thing. There's a good reason for this— this is a science book and scientists the world over use SI units for consistency. Of course, in everyday life in the United States, people use what are commonly known as English units (pounds, feet, inches, miles, and the like).

The book you have in your hands, *Energy,* covers not just the basics of energy (work, kinetic energy, potential energy, and the transformation of energy), but also energy as it relates to simple machines, temperature, and heat transfer. The final chapter draws on most of the concepts presented in the rest of the book to address how we generate electricity for various purposes. I do not address a number of energy topics that you might find in a physical science textbook, choosing instead to provide just enough of the basics so you will be able to figure out those other concepts when you encounter them. You might also notice that this book is not laid out the way these topics might be addressed in a traditional high school or college textbook. That's because this isn't a textbook. You can learn a great deal of science from this book, but it's not a traditional approach.

One more thing to keep in mind: You actually CAN understand science. It's not that hard when you take it slowly and don't try to jam too many abstract ideas down your throat. Jamming things down your throat, by the way, seemed to be the philosophy behind just about every science course I ever took. Here's hoping this series doesn't continue that tradition.

Acknowledgments

The Stop Faking It! series of books is produced by the NSTA Press: Claire Reinburg, director; Carol Duval, project editor; Linda Olliver, art director; Catherine Lorrain-Hale, production director. Linda Olliver designed the cover from an illustration provided by artist Brian Diskin, who also created the inside illustrations.

This book was reviewed by Lynn Cimino-Hurt (Flint Hill School, Virginia); Olaf Jacobsen (Mesa Public Schools, Arizona); and Daryl Taylor (Williamstown High School, New Jersey).

About the Author

As the author of NSTA Press's *Stop Faking It!* series, Bill Robertson believes science can be both accessible and fun—if it's presented so that people can readily understand it. Robertson is a science education writer, reviews and edits science materials, and frequently conducts inservice teacher workshops as well as seminars at NSTA conventions. He has also taught physics and developed K–12 science curricula, teacher materials, and award-winning science kits. He earned a master's degree in physics from the University of Illinois and a PhD in science education from the University of Colorado.

About the Illustrator

The recently-out-of-debt, soon-to-be-famous, humorous illustrator Brian Diskin grew up outside of Chicago. He graduated from Northern Illinois University with a degree in commercial illustration, after which he taught himself cartooning. His art has appeared in many books, including *The Beerbellie Diet* and *How a Real Locomotive Works.* You can also find his art in newspapers, on greeting cards, on T-shirts, and on refrigerators. At any given time he can be found teaching watercolors and cartooning, and hopefully working on his ever-expanding series of *Stop Faking It!* books. You can view his work at *www.briandiskin.com.*

Materials List

The following list is for the workshops, and the quantities listed are for two people. For classroom use, fewer materials might be necessary for groups of three or four students.

- 2 balloons
- 2 rubber bands
- 2 magnets (any kind)
- 1 battery (any kind)
- 1 match
- 2 short lengths of string
- 1 long (1–2 feet) length of string
- 2 metal washers
- 3 marbles (same size and weight)
- 1 ruler with groove down the middle
- 1 bicycle wheel
- 1 paper cup cut in half lengthwise
- 2 sections of Hot Wheels track with connector
- 1 index card
- 1 pair of scissors
- 2 sewing machine bobbins (metal) or 2 plastic pulleys (from science supply outlet)
- 1 set of plastic gears (not crucial, but nice to have if available)
- 1 alcohol thermometer
- 1 empty plastic 1-liter or 2-liter bottle
- 1 bimetallic strip (not crucial; available through science supply outlet)
- 1 business card
- Crayon shavings
- 1 votive candle
- 1 sheet of white construction paper
- 1 sheet of black construction paper
- 1 table lamp
- 1 10-foot length of insulated bell wire
- 1 magnetic compass
- 1 toilet paper tube
- 1 quality bar magnet

Recognizing Energy

Energy is such a common idea that it might seem silly to have a chapter that's all about recognizing it. After all, we talk about energy all the time. Should you buy energy-efficient windows? The country needs an energy policy. That little kid at the store who's screaming at the top of her lungs sure has a lot of energy. Candy bars are good for an energy boost. Close the refrigerator; you're wasting energy!

We use the word energy a lot, but can you come up with a quick and easy definition that you didn't find in a textbook? Can you hold energy in your hands?

Things to do before you read the science stuff

To help you with the little dilemma I just posed, I want you to do the following things. Some are actual activities and some are questions to answer. If you spend a bit of time on these, then the section that follows will make more sense.

- Roll a marble or ball across the floor. Does this marble have energy while it's rolling? How do you know?
- Does the wind have energy? How do you know?
- Clap your hands. Any energy present when you do that?

- Hold an unstretched rubber band in your hand. Does it have any energy? Now stretch the rubber band. Any energy now?

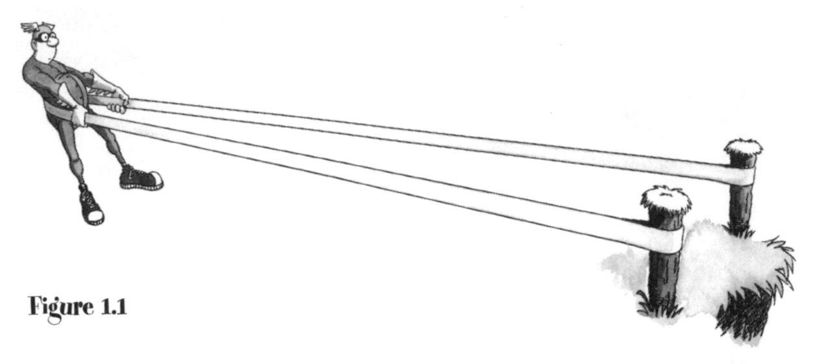

Figure 1.1

- Take your marble or ball and put it on the floor, at rest. Now pick it up and put it on a table. Push it off the table so it falls back to the floor. In which of these situations (on the floor, on the table, falling to the floor) did the ball have energy? Did it have more energy in one situation than in another?

Figure 1.2

- Grab a couple of small magnets. Refrigerator magnets will do if you don't have anything else lying around (tell the kids you'll put their artwork back when you're done). Place the magnets next to each other so they stick together. Now pull them apart just a tiny bit (about a centimeter). Let go—they should jump back together (Figure 1.2).

  In which of the three situations (stuck together, pulled apart, jumping back together) did the magnets have energy? Any more energy in one situation than in another?

- Hold an unlit match in your hand. Does the match have energy? Strike the match. Any energy now? How do you know?
- Does a battery have energy? Does the Sun have energy? In both cases, how do you know?

The science stuff

Now that you've done all those things, I can reveal that most of the questions in the previous section were trick questions. *Everything* I listed had energy. If that doesn't fit with your answers, be patient and I'll address all the situations by the end of this section. We'll start with the most obvious, though.

Anything that's moving has a special kind of energy known as "moving energy." To make that sound less silly and more sophisticated, physicists use a derivation of the Greek word *kinetikos* (which not surprisingly means "moving") and call this kind of energy **kinetic energy**. Later on, we'll figure out how to calculate the kinetic energy of something. For now, just know that a rolling ball, a falling ball, a running dog, two magnets heading towards each other, and anything else in motion, all have kinetic energy.

Topic: energy

Go to: *www.sciLINKS.org*

Code: SFE01

Topic: kinetic energy

Go to: *www.sciLINKS.org*

Code: SFE02

Topic: potential energy

Go to: *www.sciLINKS.org*

Code: SFE03

What about the wind? Does it have kinetic energy? Sure, as long as you believe in that invisible stuff called air. Wind is nothing but moving air.

On to the next kind of energy. Chances are you said a stretched rubber band has energy. How do you know? Because when you let go of it, it snaps back into place and perchance flies away and hits a bystander, who feels the result of that energy. By the same token, the two magnets, when pulled apart, have energy. You know this because they jump back together when you let go. The energy these things have is an energy of *relative position* (the position of one object compared to the position of another object) or shape. The rubber band has energy because it's stretched rather than relaxed. The magnets have energy because they're apart rather than together.

Figure 1.3

In this relative position, the magnets have no potential energy

In this relative position, the magnets have potential energy

When something, or a group of things, has energy simply because of relative position or shape, it's known as **potential energy**. Showing remarkable consistency, physicists chose this name because, like kinetic energy, it's derived from a Greek word. Ummm—okay, no they didn't. The word *potential* actually comes from a Latin word, but the meaning sort of fits. If someone has potential, they have the ability to do something, even if they never get around to doing it. A stretched rubber band has potential energy and thus the ability to do something—snap back into place if you let go. Two separated magnets have potential energy and the ability to jump back together when you let them go.

Of course, the "ability to do something" isn't associated only with potential energy. A rock with kinetic energy has the ability to break a window, and a roller coaster with kinetic energy has the ability to turn your knuckles white. (Okay, strike that last one. Actual energy considerations with roller coasters coming up in Chapter 2.)

How about when you lift a ball and place it on a table? Does the ball have potential energy once it's on the table? Yes and no. You did create a situation where the ball had the ability to fall to the floor if knocked off the table, but that potential energy was created not just in the ball but in the ball and the Earth! You see, the reason the ball falls is because there's a force of *gravity* between the Earth and the ball. If the Earth weren't around, the ball wouldn't fall from the table to the floor. So the potential energy resides in the pair—Earth plus ball—just as the potential energy of separated magnets lies in the pair of magnets and not just in one of them.

Figure 1.4

The separation of Earth and ball gives this combination potential energy

A quick review before we move on. We've identified two kinds of energy. The first is kinetic energy, which is energy of motion. The second is potential energy, which is the energy two or more things have due to their relative position or shape. One caution regarding potential energy. Some books will refer to potential energy as "stored energy." Makes sense, because things that have potential energy and no kinetic energy are not moving, so the energy must be "stored." However, not all energy that is stored is potential energy. Imagine holding a bicycle just off the ground with its wheels spinning away (Figure 1.5).

Figure 1.5

What will happen if you drop this bike to the ground? Well, it will probably fall over but before it does that it should move forward a bit, exhibiting kinetic energy (Figure 1.6).

Figure 1.6

It wouldn't be too far-fetched to say that the bike had "stored energy" when it was off the ground with its wheels spinning. That kinetic energy (spinning wheels) isn't accomplishing anything—it's just there. You can think of it as stored energy. What this means is that, although all potential energy can be thought of as stored energy, not all stored energy is potential energy.

All right, let's press on to all those other things I had you do. Start with clapping hands. Any energy? Well sure, your hands have kinetic energy while they're coming together, but there's also sound. Does sound have energy? It might not be obvious but yes, sound is a form of energy. If there wasn't any energy in sound, how could it cause your eardrum to move?

I'm betting you said the lit match had energy. After all, you can feel the heat from it. You could call that *heat energy,* but I'm going to suggest a different term—*thermal energy*. There's a reason for making that distinction and it has to do with a specific definition of what heat is. I'll cover that in a later chapter. If you think a lit match has energy, then you probably also think the Sun has energy (big match).[1] How about an unlit match? If an unlit match doesn't have energy, how do you get fire from it? Certainly all that energy of the fire doesn't come from just striking it. Otherwise any piece of wood would catch fire when you strike it. The key with an unlit match is that it has a specific arrangement of chemicals in the tip. We call that arrangement *chemical energy*. And while we're talking about chemical energy, that's what's in a battery (as long as it's not dead). A specific arrangement of chemicals inside the battery gives it its energy.

By now you might get the idea that we could go on naming different kinds of energy forever. Maybe we can't go on forever but we can name many different kinds of energy: elastic energy, thermal energy, radiant energy, electrical energy, chemical energy, nuclear energy. Seems complicated, yes? Not really, because it

[1] Actually, the Sun has a wee bit more energy than a big match. There are nuclear reactions (yep, like hydrogen bombs) taking place inside the Sun, so you could say the Sun has nuclear energy.

turns out that just about every single kind of energy boils down to two kinds—kinetic energy and potential energy. That's fortunate, because physicists like things to be simple. The best theories of how the universe behaves tend to be the simplest ones, and it's a sure bet that when your scientific explanation gets really, really complicated, you're on the wrong track.

SCiLINKS. THE WORLD'S A CLICK AWAY

Topic: forms of energy

Go to: *www.sciLINKS.org*

Code: SFE04

One big exception to kinds of energy being reduced to kinetic or potential energy is something called *mass energy*. You can think of mass as being the "amount of stuff" in an object—elephants have lots of mass and tsetse flies have very little mass. At any rate, it turns out that all things that have mass have energy just because of that mass. A really smart guy named Albert Einstein figured that out, and it's represented by that very famous equation $E = mc^2$. (In this equation, E stands for energy, m stands for mass, and c stands for the speed of light.) We won't be doing anything else with mass energy in this book, so now that I've introduced it, you can forget it if you want. But now at least you know why I told you at the beginning of this section that everything you looked at had energy.

A final thought before you hit the next section. I've shown you lots of ways to recognize energy, but you'd still have a tough time filling the blank in "energy is _____." Sure, you could say something like "energy is cute," but what I'm talking about is a definition. We don't have a definition yet, nor a clear picture of what energy *is*. That's because, common as it is, energy is a pretty abstract concept. Not so abstract, though, that you can't get a grasp on it. If I thought otherwise, I'd end the book right here!

More things to do before you read more science stuff

Figure 1.7

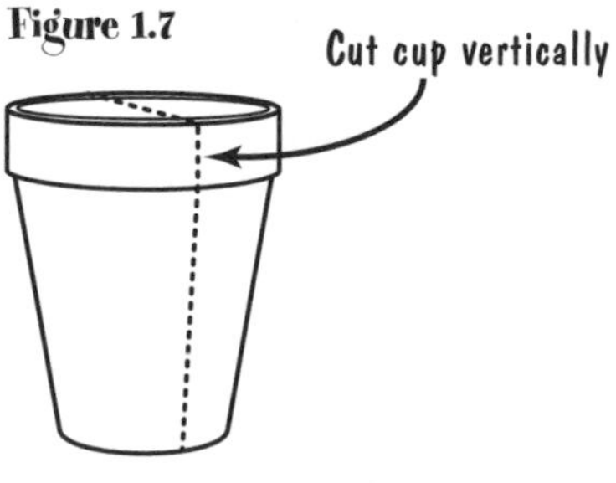

Cut cup vertically

Get a paper or disposable plastic cup and a couple of marbles or small balls. Cut the cup in half vertically, as shown in Figure 1.7.

Place one of the half-cups on its side on a smooth, level surface such as a linoleum floor or a countertop, as in Figure 1.8. It should look like a tunnel that's closed off at one end.

Figure 1.8

Take a marble and roll it toward the open end of your neat little nonfunctional tunnel. When the marble hits the back, it should push the half-cup along the surface a short way. Do this a few times, rolling the marble at differing speeds. The faster you roll the marble, the farther the cup should slide before coming to rest. Don't expect the cup to slide in one smooth motion; usually it goes a short distance when the marble first hits it and then slides again as the marble catches up and hits it again.

Now find something you can use as a ramp for the marbles to roll down. A clipboard or record album cover works well (you younger people ask your parents what that might be), as does one of those plastic rulers with a groove down the center on which the marble can roll. Prop your ramp against a box, a stack of books, or something similar, so the ramp is at about a 30° angle (this angle isn't critical, so don't measure it). And yes, you should do this on that smooth, level surface you've been using (Figure 1.9).

For what you're about to do, it's important that the ramp stay in the same position throughout. So even though the particular angle you use isn't important, that angle should stay constant. Sounds like a job for duct tape.

Place your half cup at the bottom of the ramp so that a marble rolling down the ramp will enter the cup and push it a ways. Set a marble about halfway up the ramp, let it go, and make sure it actually enters the cup.

Experiment time. With the cup positioned right at the bottom of the ramp, place the marble 1/3 of the way up the ramp. Let it go and then mark how far the cup goes before it stops (Figure 1.10).

Figure 1.9

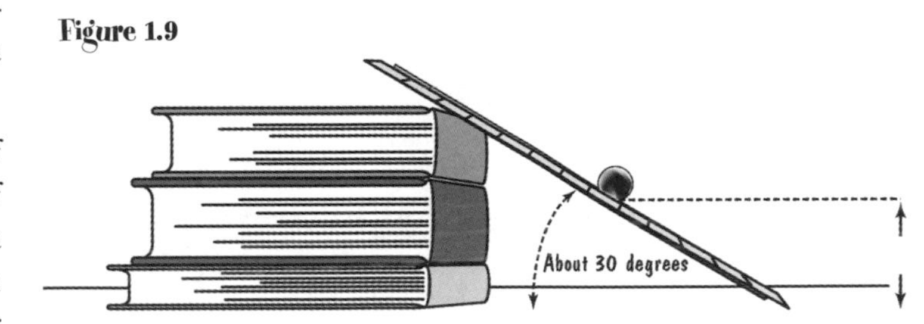

Figure 1.10

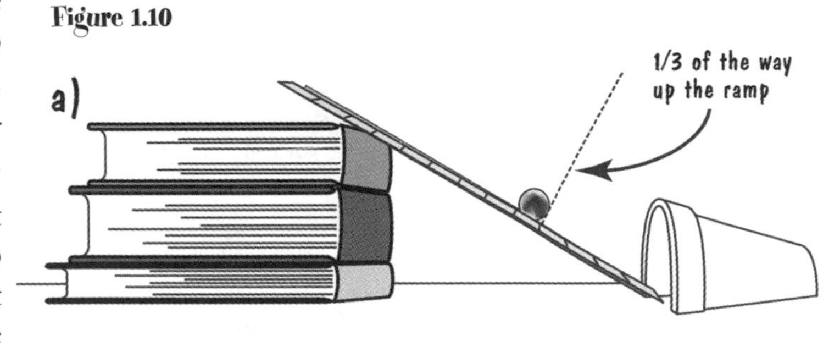

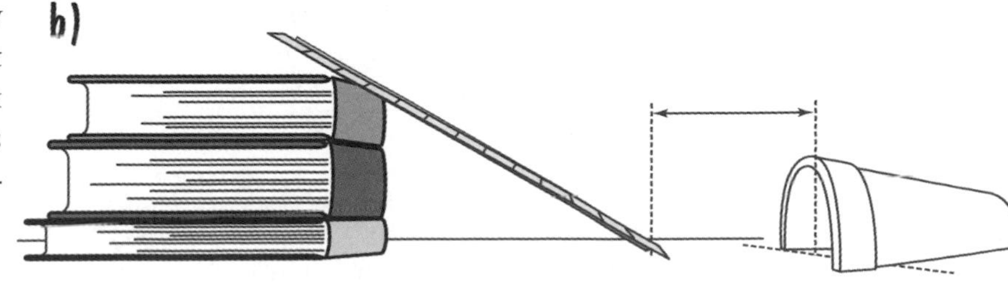

Measure distance the cup moved

Obviously you need to be careful about marking the distance the cup moves; use a small piece of paper or something similar rather than a permanent marker. Also, the cup usually twists a bit, so you should determine the distance as shown in Figure 1.11.

Figure 1.11

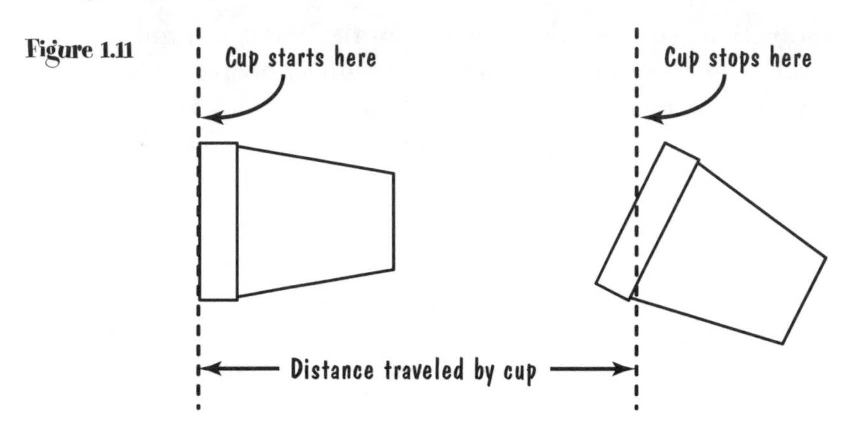

Distance the cup travels is measured at the center of the lip of the cup

Repeat what you just did several times until you consistently get about the same distance moved (expect small differences each time—that's normal). Now repeat everything you've done, except with the marble $^2/_3$ of the way up the ramp. Again, do this a number of times until you consistently get about the same distance moved by the cup. Compare the distance the cup moves when you release the marble $^1/_3$ of the way up the ramp with the distance the cup moves when you release the marble $^2/_3$ of the way up the ramp.

Repeat again with the marble all the way at the top of the ramp. After doing that, you have three distances to compare—the distances the cup moves when the marble is released $^1/_3$ of the way up, $^2/_3$ of the way up, and at the top of the ramp.

Because this little activity is so much fun, why not take one more measurement? Repeat your measurements for $^2/_3$ of the way up the ramp, using two marbles instead of one. Make sure the marbles are about the same size and seem to weigh about the same. If you're using a ruler, you'll have to place the marbles one behind the other. If you're using some other ramp, you can place them side by side, making sure both of them enter the cup at the bottom. Compare the distance the cup moves using two marbles with the distance the cup moves using one marble. If you're feeling ambitious, you can repeat using three marbles, although it can be difficult to get all three to go into the cup.

Okay, I lied. One more thing to do. Compare the distances the cup moves using one marble in two different situations:

(a) Ramp at the set angle, marble halfway up the ramp

(b) Ramp set at a different angle, marble wherever it has to be on the ramp so its vertical height above the surface is the same as in Figure 1.12a

Seeing as how that probably totally confused you, take a look at Figure 1.12.

Figure 1.12

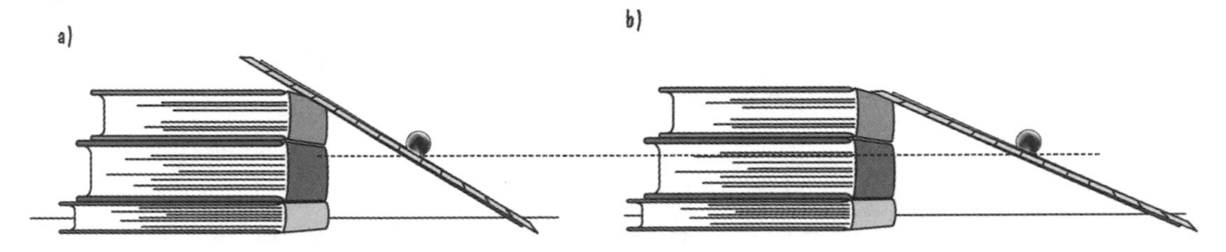

More science stuff

We'll start with a big assumption, which is that the more energy the marble has, the farther it moves the cup. To take it a step further, I'm going to claim that if the marble moves the cup twice as far, the marble has twice the energy. If it moves the cup three times as far, it has three times the energy. For reasons I'll explain in the next chapter, this assumption turns out to be a pretty good one. It also just plain makes sense. Something with twice the energy should have about twice the effect on something else.

Unless something really strange happened, you should have gotten the following results in your experiment.

- One marble released $^2/_3$ of the way up the ramp pushes the cup twice as far as one marble released $^1/_3$ of the way up the ramp.
- One marble released from the top of the ramp pushes the cup three times as far as one marble released $^1/_3$ of the way up the ramp.
- For a given distance up the ramp, two marbles push the cup about twice as far as one marble, and three marbles push the cup about three times as far as one marble.
- When you keep the *vertical* height of the marble the same, regardless of the angle of the ramp, the marble pushes the cup about the same distance.

If you got completely different results from these, you might want to check your procedure. Did you keep the ramp in the same position throughout? Did you measure the distance as shown in Figure 1.11? Was the surface really smooth and level, or were there bumps that might mess things up? Whatever your answers, you have the option of either redoing the steps or just taking my results.

Of course, even if you did everything just right, you probably didn't get exactly my results. We're not looking to publish your experimental results but rather to give you a general idea of how marbles and ramps behave.

Okay, what in the world do all these results mean? First let's concentrate on the height. Releasing the marble from twice the height gives it twice the energy at the bottom (as evidenced by the cup moving twice as far). Releasing the marble from three times the height gives it three times the energy at the bottom. And remember that it's the *vertical* height that matters. You saw that when you changed the angle of the ramp and kept the vertical height the same.

Now comes a big leap of faith. I claim that this relationship between height and energy applies to *everything* and not just marbles that roll down ramps. In fact, what we're talking about is **gravitational potential energy**, the potential energy everything has because of its height above the Earth's surface. An object (actually the combination of the object and the Earth) has gravitational energy that depends on the separation (height of object) between the object and the Earth.

The height, however, isn't the only thing that affects gravitational potential energy. Recall that, for a given height, two marbles had twice the energy of one marble and three marbles had three times the energy of one marble. Because the marbles were all about the same size and weight, they all had approximately the same *mass*. That means that two marbles have twice the mass of one marble, and three marbles have three times the mass of one marble.

The bottom line here is that gravitational potential energy depends not only on the height, but also on the mass of the object. The formula looks like this:

$$\text{Gravitational potential energy} = mgh$$

where m is the mass of the object, h is the height of the object above some surface, and g is a special number that equals 9.8 meters per second squared.[2]

> Note to those of you who cringe when you see a formula like this: The letters are called *variables* and they're placeholders for actual numbers.
>
> Also, when two or more letters are put side by side, that means you're supposed to multiply them together. For example, the expression vt, where v equals 3 and t equals 7, is equal to 3 times 7, or 21.
>
> So, to figure out an actual number for gravitational potential energy, you put numbers in for m, g, and h, and multiply all three together.

If you use SI units for m (kilograms), g (meters per second squared) and h (meters), the unit that results for energy is known as the *joule*, named after Sir

---

[2] g is actually the acceleration due to gravity of objects that are falling freely near the Earth's surface. For a more thorough understanding of what g represents, check out the *Force and Motion* book in this series.

James Joule, an amateur physicist and son of an English brewer (Gotta like a guy like that!). The surname is pronounced *jowl*, but for some reason, physicists pronounce the unit *jool*. To see that this formula fits with our experiment, put in any old numbers for m and h (use 9.8 meters per second squared for g), say 4 kilograms and 6 meters. That gives us an energy of:

SCILINKS® THE WORLD'S A CLICK AWAY

Topic: James Prescott Joule
Go to: *www.sciLINKS.org*
Code: SFE05

$$\text{Gravitational potential energy} = mgh$$
$$= (4\ \text{kg})(9.8\ \text{m/s}^2)(6\ \text{m})$$
$$= 235\ \text{joules (approximately)}$$

If you double or triple the 4 kg to 8 kg or 12 kg, you get twice or three times the energy. If you double or triple the 6 m to 12 m or 18 m, you get twice or three times the energy. Those results agree with what we found in our experiment. In Figure 1.13, George has twice the gravitational potential energy of Wally, Martha has six times the gravitational potential energy of Wally, and Petunia has three times the gravitational potential energy of Wally. Hope that makes sense.

Before moving on, there's one thing that might be bothering you. If not, it should bother you after I mention it. Maybe you did your experiment on the second floor of a house, or some other place that wasn't level with the surface of the Earth. Yet the values for h were measured from the surface the cup was sliding on, rather than from the surface of the Earth. Shouldn't something sitting on the second floor of a building automatically have more gravitational potential energy than something sitting on the first floor of a building?

Figure 1.13

Well, yes. In fact, to get an accurate number for the gravitational potential energy between a marble and the Earth, we would have to use a much more complicated formula and we would have to use the distance between the marble and the center of the Earth! It turns out, however, that we can get a good picture of what's going on by considering only *changes* in potential energy. The *change* in potential energy in going from the bottom of a ramp to halfway up is the same on the first floor as it is on the second floor as it is on the tenth floor.

So, I've been lying just a wee bit. The formula *mgh* doesn't give you *the* potential energy between an object and the Earth, but if you calculate it at one height and then at another height, it *does* give you the right answer for the change in potential energy that took place in going from one height to the next. You can arbitrarily choose a point from which to measure all the heights (known as the *reference level*), and everything works out just fine. If that's still a bit fuzzy, don't worry. I'll hit on it again in the next chapter.

Even more things to do before you read even more science stuff

Back to the ramp and marbles (no need for the cup this time). Set up the ramp as you did at the start of these experiments and mark off a distance of about 1 meter (the exact distance isn't important) from the base of the ramp. (See Figure 1.14.) You're going to time how long it takes for the marble to go from the base of the ramp to the point you mark off, so a stopwatch will be a big help. No big deal if you don't have a stopwatch, though. Just use the old "one one thousand, two one thousand . . ." method to get a rough idea of the time in seconds.

Figure 1.14

Figure 1.15

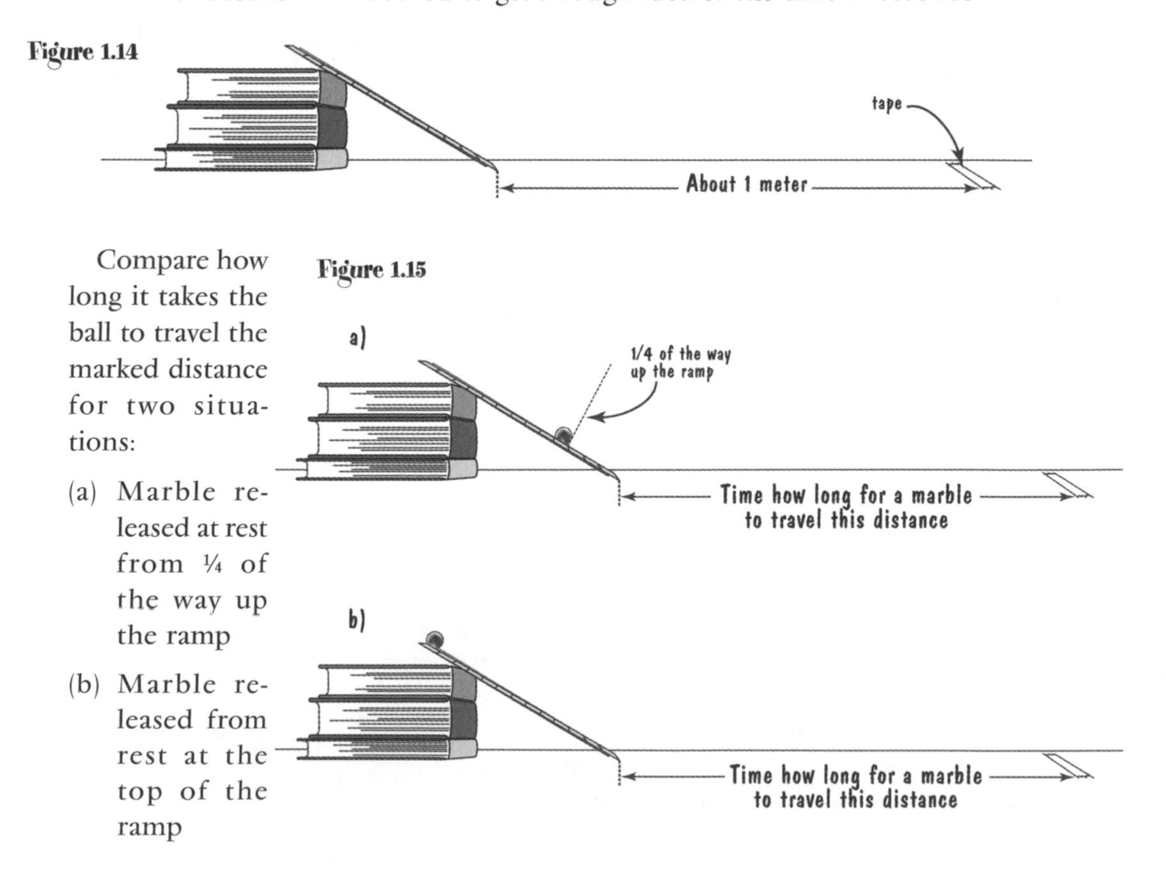

Compare how long it takes the ball to travel the marked distance for two situations:

(a) Marble released at rest from ¼ of the way up the ramp

(b) Marble released from rest at the top of the ramp

Do this a bunch of times until you get more or less consistent results. Remember, this ain't rocket science.

Even more science stuff

So you know where I'm headed in this section, I'm going to do something similar to what I did in the last explanation section. I'm going to use the results of the activity you just did to try and convince you of the plausibility of a formula for the kinetic energy of an object.

When I did the previous experiment, I got to just a bit more than "three one thousand" with the marble ¼ of the way up the ramp and I got to "two one th . . ." with the marble at the top of the ramp. This tells me it took the marble just about twice as long in the first situation as in the second situation. Hopefully you got similar results. With a good stopwatch, I'm betting you got almost exactly double the time.

What this result means is that the marble is moving twice as fast in the second situation (released from the top of the ramp) as in the first situation (released from ¼ of the way up the ramp). Stands to reason. If you take half the time to go a distance, you must be moving twice as fast. If you don't believe that, try it in a car sometime, where you can check your speed the whole way. What's one little speeding ticket in the name of science?

Now think back to moving the cup with the marble. There we found out that a ball released from the top of the ramp would have *four times* as much energy as one released from ¼ of the way up the ramp. That tells us that when a marble is moving twice as fast, it has four times as much energy.

Figure 1.16

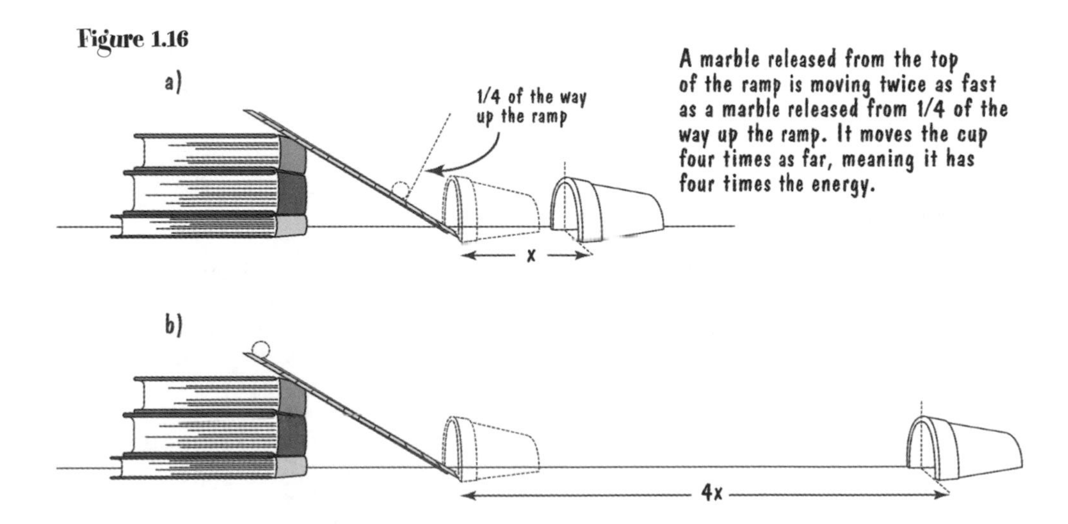

This suggests that maybe the kinetic energy of an object depends on the *square* of the speed, because $2^2 = 4$. Of course to check that out, we'd have to repeat the experiment with the marble released, say, $1/9$ of the way up the ramp and then from the top of the ramp. Then the marble would have nine times the energy at the top as it would $1/9$ the way up the ramp. If our "squared" relationship is true, then we would get speeds at the bottom that differed by a factor of 3, because $3^2 = 9$. In fact, if you do that carefully, that's the result you get. Feel free to try that in your spare time!

We also know that kinetic energy depends on the mass of an object, because if we double the mass (two marbles) and keep the speed the same, the cup will move twice as far (you actually did that earlier in the chapter). All right, enough beating around the bush. Here's the formula for kinetic energy:

$$\text{Kinetic energy} = \frac{1}{2} mv^2$$

where m is the mass of the object and v stands for velocity.

There is a definite difference between speed and velocity, but for now you can think of them as being the same thing. As for the ½, well, it's just there. I'll try to justify that number in the next chapter. I know you must be getting pretty excited about the next chapter given all the things I'm putting off until then! Notice that kinetic energy depends on mass and on the *square* of the velocity, as we knew it had to.

Now we have expressions for two important kinds of energy—gravitational potential energy or mgh and kinetic energy or $\frac{1}{2} mv^2$. If an object is a given distance above a reference point, you can put that distance in for h; plug in the object's mass, m; plug in 9.8 for g; and get a number that represents the object's gravitational potential energy. If an object is moving at a given velocity, you can square that velocity, multiply by the object's mass, multiply by ½, and get a number that represents the kinetic energy of the object.

That bit of knowledge might not have you dancing around the room, but we can use those numbers to solve quite a few real-life problems, as you'll see in the next chapter. There I go again. Maybe it's time to get to the next chapter.

Chapter Summary

- Energy can take on many different forms, such as thermal energy, sound energy, electrical energy, and chemical energy.
- With the exception of mass energy, all forms of energy are some kind of kinetic or potential energy.
- Kinetic energy is the energy something has by virtue of its motion. It equals $\frac{1}{2} mv^2$, where m equals the mass of the object and v is its velocity.

- Potential energy is the energy an object or group of objects has due to the position or shape of the object(s). Potential energy often can be thought of as "stored energy," but it is incorrect to say that all stored energy is potential energy.
- Gravitational potential energy is the potential energy contained in the system consisting of an object and the Earth. For objects near the surface of the Earth, we often speak of the *object* having the gravitational potential energy, even though that energy resides in both the object and the Earth. For objects near the surface of the Earth, gravitational potential energy is equal to mgh, where m is the mass of the object, g is equal to 9.8 meters/sec^2, and h is the height of the object above some arbitrarily chosen reference level.

Applications

1. Let's see how well you understand the process we went through in this chapter. Get a rubber band and stretch it. It now has potential energy, right? Yep, because it's an energy that results from the new shape of the rubber band, not because it's moving. Now here's the challenge. Use the half-cup and a smooth, level surface to figure out how the potential energy in a stretched rubber band depends on the distance it's stretched. Cue the Jeopardy theme song while you think about it . . . .

   Did you figure out how to do it? If not, here's the answer. Stretch the rubber band as if you're going to fire it at someone. Note the position of the end of the rubber band before it's stretched and then after it's stretched (Figure 1.17).

   Fire the rubber band at the half-cup so you hit the back side of the cup and move it. Notice how far the cup goes (Figure 1.18).

Figure 1.17

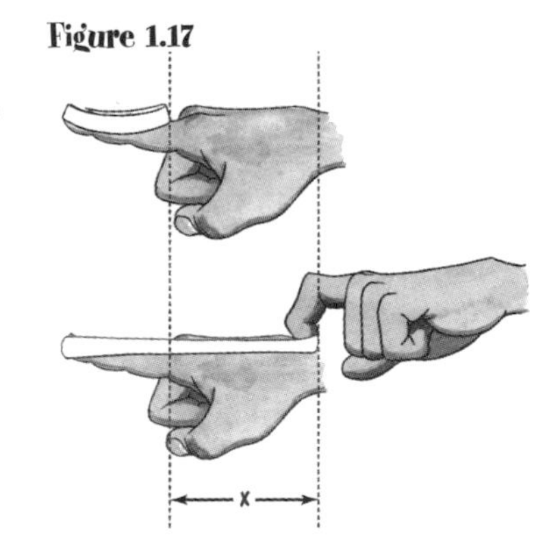

Figure 1.18

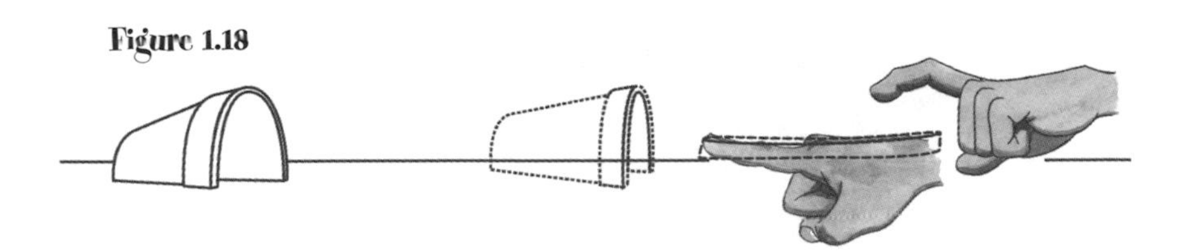

Figure 1.19

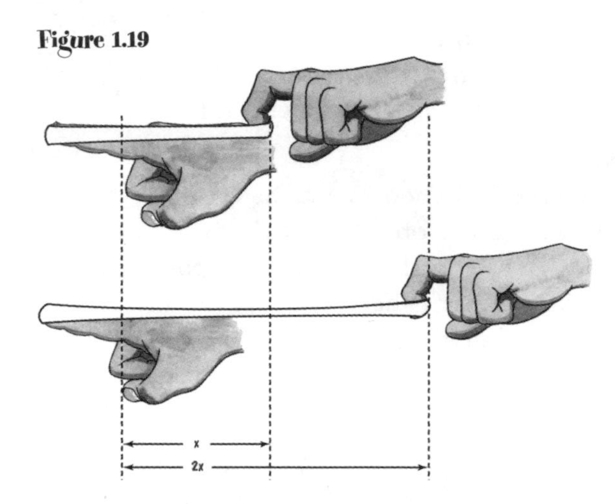

Now stretch the rubber band twice as far and repeat (Figure 1.19).

Here's the result I get: Stretching the rubber band twice as far moves the cup about four times as far, meaning the potential energy of the rubber band depends on the *square* of the stretched distance.

2. Using some special relationships known as *kinematic equations,*[3] you can calculate how fast something will be moving when it falls from rest through a certain distance.

Let's say a rock with a mass of 2 kilograms falls off a cliff 20 meters high. Using those kinematic equations, I can calculate that the rock's velocity just before it hits is 19.8 meters per second. Your task: Calculate the gravitational potential energy of the rock (use h = 20 meters) at the top of the cliff. Then calculate the kinetic energy of the rock, just before it hits bottom. You should get the same number for each kind of energy. I'll explain this result in the next chapter. Hmmmm—foreshadowing for the science geek!

[3] See the *Force and Motion* book for a brief discussion of these.

Energy on the Move

Up until now, we've been talking about things *having* a certain amount of energy and not about things gaining, losing, or changing their form of energy. We've already seen how energy can change, though. A marble at rest at the top of a ramp has a certain amount of gravitational potential energy and no kinetic energy. After you let go, and the marble reaches the bottom of the ramp, the marble has less gravitational potential energy and it has acquired some kinetic energy (it's moving now!). Seems logical that what happened was that the gravitational potential energy transformed into kinetic energy. Energy transformations like this take place all the time. This chapter deals with keeping track of those transformations, and we'll end up with an incredibly useful principle for, among other things, building amusement parks. What better use for science?

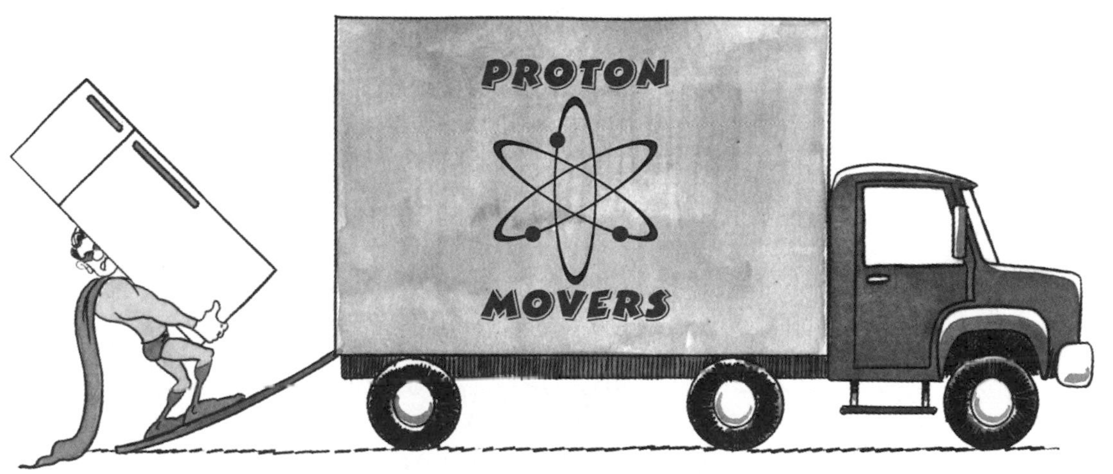

"Hmph! It said 'Energyman. Man *on* the move!' not '*for* the move!' I really neet to get an agent."

Things to do before you read the science stuff

Below are a number of simple activities to perform. As you do each one, try to track any energy transformations that take place. What different kinds of energy (kinetic, potential, sound, light, thermal, etc.) do you observe, and what seems to be transforming into what? Of course, I'm going to provide answers in the next section, so you *could* just skip ahead and not actually do the activities. Wouldn't be as much fun, though, and you really do learn concepts better when you have physical experiences with which to connect them.

- Slide a book or pencil or just about anything across the floor. As the object first starts moving, it clearly has energy. It slides to a stop, though. Where did that initial energy go? Did it just disappear?
- Blow up a balloon but don't tie it off. Let go of the balloon. Forgot how much fun that was to do, huh? Stop and identify the different kinds of energy and the transformations.
- Grab a rubber band and a small wad of paper. Using Figure 2.1 as a guide, fire the paper wad across the room. If this brings back memories, one of the following is probably true: a) you were not the teacher's pet when you were a kid or b) you got hit in the back of the head with just such a piece of paper last week. Identify the kinds of energy and the transformations.

Figure 2.1

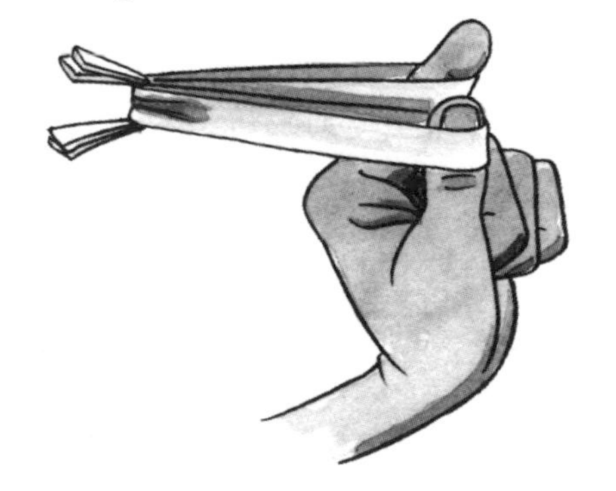

- Get a couple of toy cars and slam them into each other. Actually, they don't have to be toy cars; you can use almost anything. I just want you to model a real head-on collision between two cars. Where does all the energy go when they collide and eventually come to a halt?
- Tie a metal washer to a piece of string about 50 centimeters long (actual length not a big deal). If you don't have a metal washer, substitute anything (a roll of masking tape, a key) to which you can tie the string and which is small enough for you to swing around (Figure 2.2). You now have a makeshift *pendulum*.

Figure 2.2

- Hold the free end with one hand and start the pendulum swinging with the other (Figure 2.3).

  What energy transformations are taking place? If you wait long enough, the pendulum will eventually slow down and then stop altogether. What happened to the energy?

- Make a second pendulum that's the same length as the first one (identical lengths are critical for this). Cut a third piece of string that's at least as long as the first two. Tie the two pendulums to the third string as shown in Figure 2.4.

  Find a helper. You and your lovely (or handsome) assistant should hold the third string horizontal and taut, with the two pendulums hanging down. Pull *one* of the pendulums to the side and let it go. Watch for a while and see what happens (Figure 2.5).

Figure 2.3

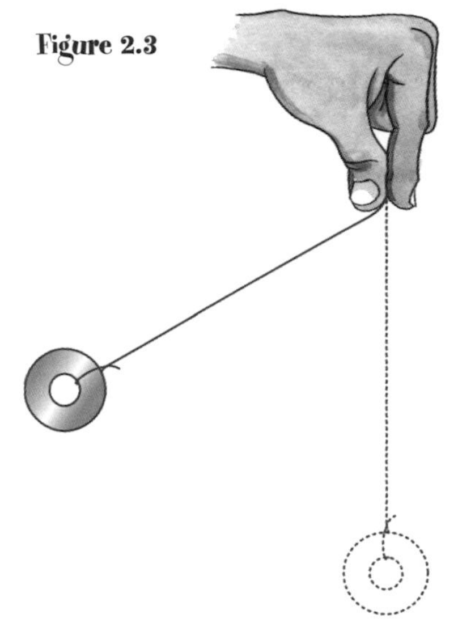

Figure 2.4

Figure 2.5

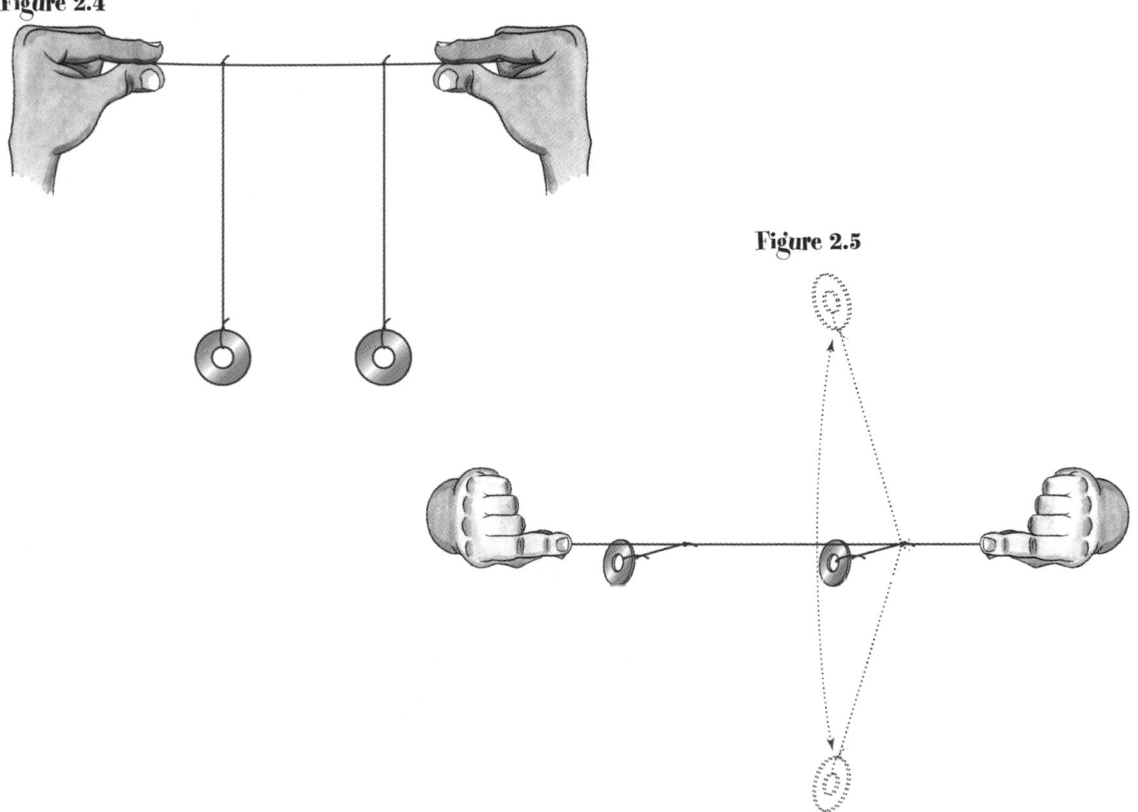

The science stuff

A sliding book has kinetic energy at the start. Then the energy goes—where? First of all, the energy does *not* just disappear. In fact, as far as we know, energy *never* disappears. More on that later. We can figure out where the book's energy went by first figuring out why it slowed down and eventually stopped. The culprit is the force of *friction* between the book and the floor. When one thing slides across another, there's a force of friction between them that opposes any motion. If you push against something opposite to the direction it's moving, it slows down and eventually stops.

Fine, but where did the energy go? To answer that, rub your hands together really fast. Feel the heat? It's generated by the friction between your hands. So what happened is that the kinetic energy of the book became thermal energy in the book and in the floor. They both got a little warmer, although you'd need pretty sensitive equipment to measure the increase in temperature. On a molecular level, the molecules in the book cover and the molecules in the floor actually moved faster, so what we call thermal energy in this case is actually kinetic energy of molecules. I'll get into that in detail in a later chapter. For now, we can draw a diagram that represents the transformation of energy.

Figure 2.6

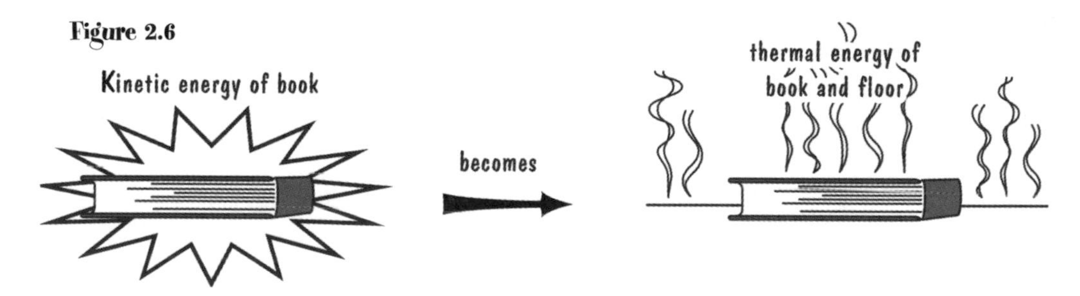

Of course, the book and floor don't stay warm. They cool off. Evidently the thermal energy of the book and floor transfers to the surrounding air. A more complete diagram is:

Figure 2.7

On to the balloon flying through the air. We start with you blowing up the balloon. The air you force into the balloon causes it to change shape by stretching, so you gave the balloon a form of potential energy (remember that potential energy is an energy of position or shape). We'll call it *elastic potential energy.* When you let go of the balloon, it gradually collapses, forcing air out the opening. The air being pushed out the opening actually pushes back on the balloon,[1] causing it to go forward. Our diagram for this situation is:

Figure 2.8

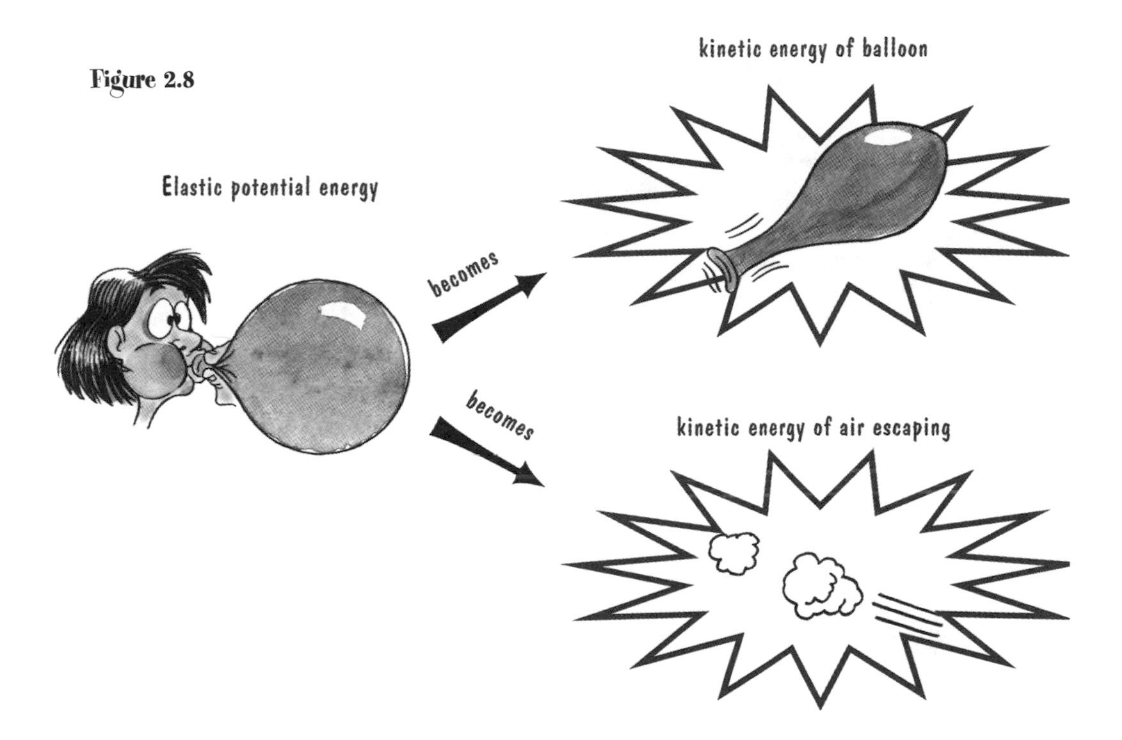

The energy transformations involved in using a rubber band to launch a wad of paper are a lot like those in blowing up a balloon and letting go. First you stretch the rubber band, giving it elastic potential energy. Then you let go. This elastic potential energy changes into kinetic energy of the paper. The rubber band moves a bit, too, so there's kinetic energy of the rubber band. And if you want to get picky, both the rubber band and the paper wad push on little air molecules, giving them a boost of kinetic energy. And of course, the energy of the paper wad actually heats up the victim of the attack when it hits him or her in the back of the neck. And just as with a book sliding against a floor, this

[1] This is an example of Newton's third law, which I'd explain here except that it takes an entire chapter to do so! Check out the *Force and Motion* book for that chapter.

thermal energy eventually ends up as thermal energy in the surrounding air. I'm also going to add in sound energy, because there's a definite sound when you let go of a stretched rubber band.

Figure 2.9

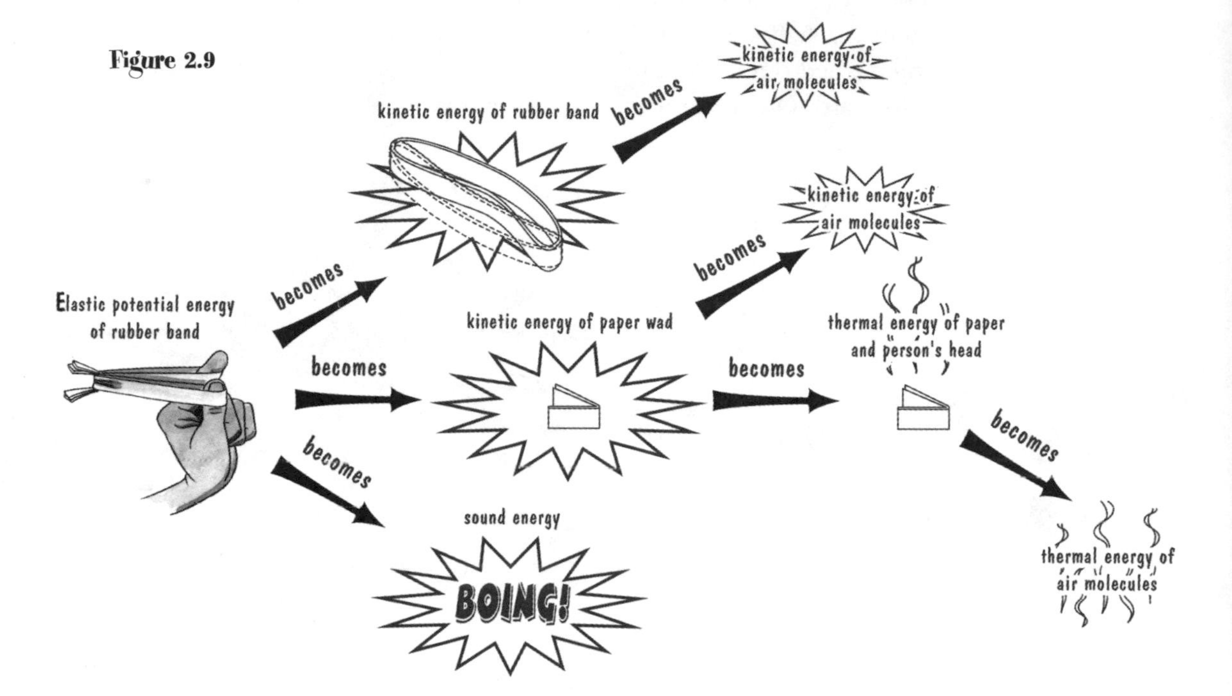

Before I go on, notice that there's a game of semantics going on (and you thought science wasn't fun and games). I can label each kind of energy in different ways (e.g., sound energy and thermal energy), but they can all be reduced to two kinds—kinetic energy and potential energy. For example, thermal energy is just kinetic energy of molecules. Sound energy also is nothing but the motion of air molecules,[2] so it's a form of kinetic energy. And although both potential energies have been elastic so far, we'll run across other forms of potential energy with different names. They'll have different names but will still be energies of position or shape.

Time for the car collision. What happens to the kinetic energy of the cars? If you've ever been near a real car crash, you'll know that it's LOUD, even at relatively low speeds. So there's lots of sound energy. There are also flying car parts, glass—you name it. Fenders and other car parts deform. After everything settles down, though, it's all at rest. That's because, due to friction, all the kinetic energies involved eventually turn into thermal energy (except for the sound, that is).

[2] See the *Sound* book in this series for a complete treatment of sound.

Figure 2.10

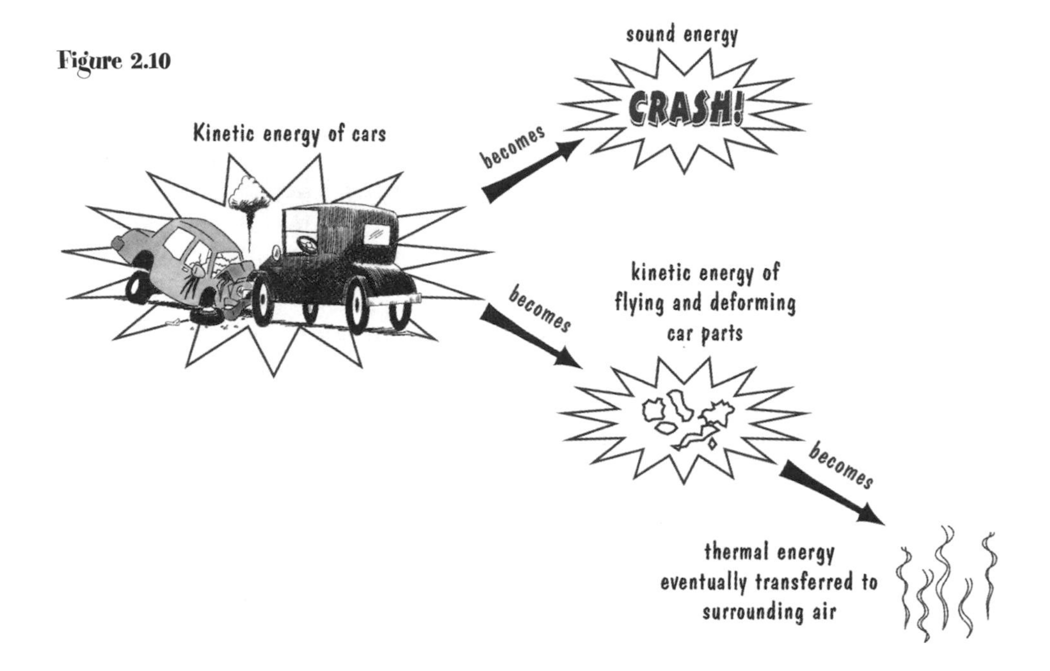

I'm going to add one more form of energy to this situation. How did the cars get their kinetic energy before the collision? There are a lot of intermediate steps, but it starts with the burning of gasoline. The explosion of gases (kinetic and thermal energy) that results from the burning of gasoline runs the engine, which, using a few gears and such in between, causes the wheels to turn. The energy for the explosion was sitting in the gasoline, in the form of chemical energy. This chemical energy is the result of certain physical arrangements of atoms and molecules so chemical energy is actually a form of potential energy. Once again, although the process of tracking energy from gasoline to crash might be complicated, all we're dealing with are different forms of kinetic and potential energy and transformations between them.

The pendulum is a bit different from the other examples because the transformations take place within the pendulum itself. When you pull it to the side, you give it gravitational potential energy, just as you give a rock gravitational potential energy when you lift it up.[3] Let it go, and that potential energy transforms gradually to kinetic energy. At the bottom of its swing, it's moving the fastest and all its energy is kinetic. That kinetic energy then transforms back into potential energy as the pendulum swings up the other side and comes to rest at

[3] I told you earlier that gravitational potential energy can't reside in just one object, such as the pendulum. It really resides in the combination of the pendulum and the Earth. From now on, though, I'll talk about a single object having gravitational potential energy, because it's just too cumbersome to clarify the object/Earth thing each time.

Figure 2.11

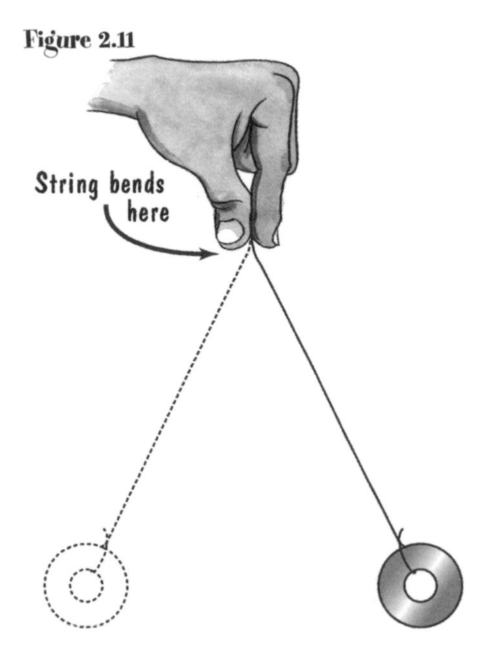

the top. Then the pendulum swings back. There's a continual transfer of kinetic energy to potential energy to kinetic energy to potential energy to—and on and on forever.

Well no, not forever. The pendulum eventually comes to a stop. Where does the energy go? The pendulum keeps hitting air molecules and giving them kinetic energy, but that turns out to be an almost insignificant source of energy loss. The important factor is the friction between the string and your fingers, and within the string itself (as the string bends, different parts of the string rub against each other). At the point where you hold the string, it bends back and forth.

To see that this bending generates thermal energy, take a paper clip or other piece of bendable wire and bend it back and forth a few times. At the spot where you're bending it, the wire gets hot. The string isn't a wire, but the same principle holds. The string gets hot at the point where it's bending back and forth. So a diagram for the pendulum energy is a simple transfer back and forth, with thermal energy siphoning off that energy bit by bit.

Figure 2.12

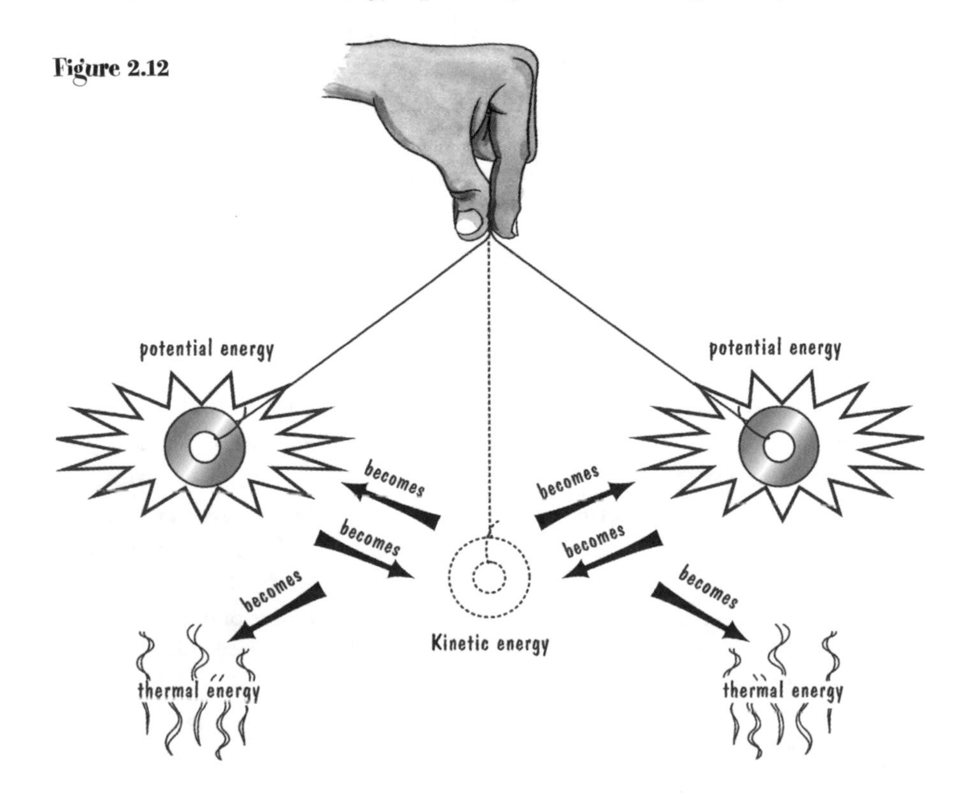

Finally, what about the double pendulum? Obviously, energy was going from one pendulum to the other and back again. Why? Because that's the way two pendulums of the same length behave! Actually, I could give you a deeper explanation, but that would take us way off track. I will tell you it has to do with something known as **resonance** (see the *Stop Faking It!* books on *Light* and *Sound*). If you want to play with the double pendulum some more, try using pendulums of slightly different length and see how that affects the motion.

More things to do before you read more science stuff

In the energy transformations so far, I've ignored one thing. In each case, you *gave* some energy to the objects involved. You pushed the book, blew up the balloon, and lifted the pendulum. Time to discuss that. First, though, do the following things. As you do, pay attention to what action you have to take to accomplish the goal.

- Give a ball or marble some kinetic energy.
- Give a ball or marble some gravitational potential energy.
- Give a spring or a rubber band some potential energy.
- Hold an object (anything!) about waist high. Walk with the object at a constant speed. Now answer this: Other than when you first got the object moving (in other words, while you were walking), did you change the object's potential energy or its kinetic energy?

More science stuff

To give an object kinetic energy, you have to hit it, kick it, push it, pull it, or shove it. Same applies to giving an object gravitational potential energy. You have to pull it or lift it or hit it so you increase the separation between the object and the Earth. Same story with giving potential energy to a rubber band or a spring. Those things don't stretch or compress all by themselves.

So, every time you exert a force[4] on an object, you give it some kind of energy, right? Wrongo, my energy transformation neophyte. Let's look at the last situation, where you carried an object along. We're going to ignore the fact that you had to pick up the object from the floor and just concentrate on your carrying the object as you walked. First, did you exert a force on it while carrying it? The answer is yes. You had to exert an upward force on the object. If you didn't, gravity would pull it to the floor. Now, did you change its energy at all?

---

[4] "Exerting a force" is a generic term that covers all hits, pulls, pushes, nudges, etc.

The speed stayed the same, so no change in kinetic energy. The height above the floor stayed the same, so no change in gravitational potential energy. No change in shape, so no change in any other kind of potential energy. So the answer is no, you didn't change its energy even though you were exerting a force on it the whole time.

Time to define something you probably didn't think needed defining—**work**. After all, you know what work is, thank you very much. But this is a special definition used only by those wacky physicists. You do work on an object when the object moves *in the direction of the force* you exert on it. The actual definition is:

Work = (force)(distance moved in the direction of the force)

In case you've never seen this notation before, putting two sets of parentheses together means you multiply the quantities inside by each other. So this is the force multiplied by the distance. We can write that relationship in shorthand notation as:

$$W = Fd_{\parallel}$$

Here the || symbol means "parallel." That's because we're interested in the distance moved in the direction of the force. Because those two are in the same direction, they're parallel. So basically this relationship says that in order to do work on something, you have to multiply the applied force by the distance moved *parallel* to the force. If you push upward on an object while it's moving horizontally, you aren't doing any work on the object. In Figure 2.13, the applied force is *perpendicular* (at right angles) to the distance moved, so no work is done.

Figure 2.13

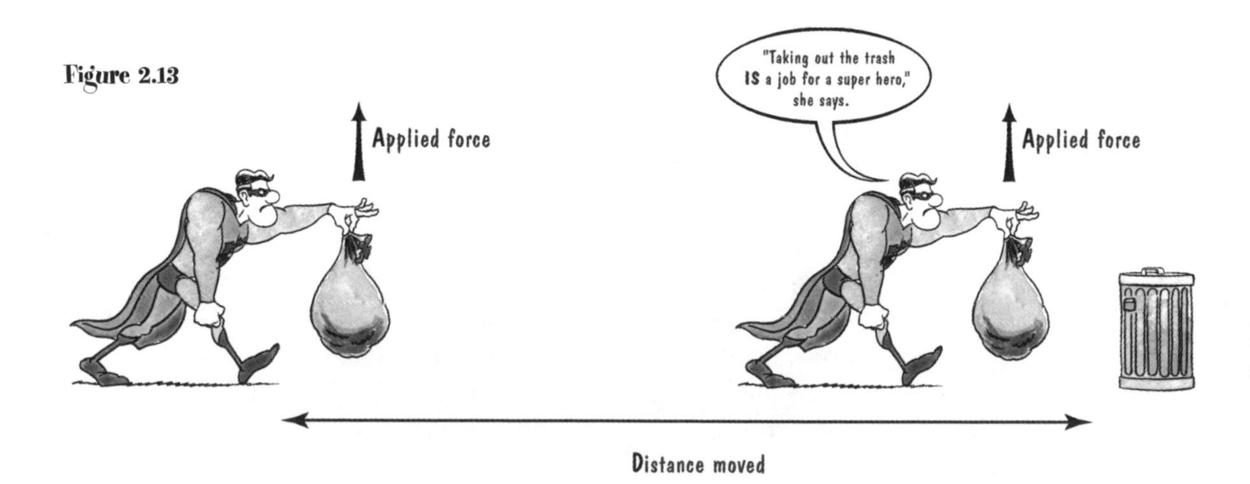

Right about now, you might be thinking that this kind of goofy definition is why you never understood science! Why in the world would scientists make up a silly definition like this? Short answer: scientists are silly. The real answer: because it turns out that with this definition of work, whenever you do work on something you change its kinetic energy. Not only that, but the amount of work you do is *numerically equal* to the change in kinetic energy. This makes work a useful concept, so useful that we give the relationship between work and kinetic energy a special name—the **work-energy theorem**. In equation form, it looks like this

Total work done on an object = change in kinetic energy of the object

Great. Now we have one made-up concept (work) equaling the change in another made up concept (kinetic energy). Actually, this is great, because it gives us some idea of what energy is. Kinetic energy is that stuff that you change when you do work on an object.

Let's try to make sense of this. When you gave a marble or a ball kinetic energy, you had to exert a force on it. Not only that, you had to exert a force in the direction of motion of the marble or ball. You did work on the ball. For kicks, let's shift to giving kinetic energy to a car (shift... get it?). We start with the car at rest, and slowly accelerate it up to a velocity v. Using force and motion relationships, you can figure out the necessary force to get an object of mass m up to a velocity v within a distance d.

Figure 2.14

You need to use a principle known as Newton's second law and a bit of (ackkk!) calculus to figure out the work applied to the car, but it can be done. The result is this:

$$\text{Work done on the car} = \frac{1}{2} mv^2$$

The expression on the right is just our previous expression for kinetic energy. That shouldn't be a surprise. The car started at rest with zero kinetic energy and ended up at velocity v with ½ mv^2 worth of energy, so ½ mv^2 is the change in the car's energy. The ½ gets in there as a result of doing the proper calculus for the situation. I told you earlier I'd let you know where the ½ came from, so there it is. It's a natural consequence of doing the math.

Okay fine. What happens when you give an object potential energy rather than kinetic energy? Does the work-energy theorem cover that? Sure, but we have to be careful about the "work" side of things. Consider taking an object with mass *m* and raising it up through a height *h,* as you would do in lifting a rock from the floor to a table (Figure 2.15). Did you do work on the rock? Of course you did. You exerted an upward force in the direction of motion.

Figure 2.15

You can exert an upward force on the rock.

h

By exerting an upward force on the rock in the direction of its motion, you do work on the rock.

Did you change the rock's kinetic energy? Nope. If we ignore the tiny bits of time when you first get the rock moving from the floor and then let it come to rest on the table, the rock does not change its velocity. That means zero change in velocity. Does that mean the work-energy theorem is garbage? After all, you did work on the rock, yet there was zero change in energy, so the work-energy theorem gives you something on the left and nothing on the right. Not good math.

To resolve our problem we have to look again at the statement of the work-energy theorem, which is

Total work done on an object = change in kinetic energy of the object

The first word—total—is the key. In considering only the force that *you* have exerted on the rock, we have not calculated the total work done on the rock. As you're lifting the rock up, gravity is pulling down on the rock. As long as you lift the rock at a constant speed, the force you exert on the rock is equal to the force gravity exerts on the rock. Because these forces are in opposite directions, the work they do on the rock cancel. Figure 2.16 shows what's going on.

Figure 2.16

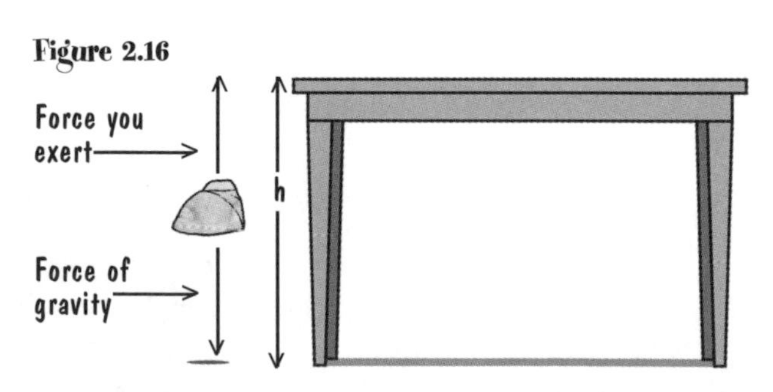

The work you do on the rock is cancelled by the work gravity does on the rock. The total work done on the rock is zero.

So, the total work done on the rock is zero, which means our work-energy theorem still works. Zero total work done on the rock and zero change in kinetic energy. We're not off the hook yet, though. When you lift a rock, don't you give it gravitational potential energy? Yep. Does the work-energy theorem explain that? Yep. Recall that the work you do on the rock is canceled by the work done on the rock by gravity. We can then say that

Total work done on the rock = work done by you – work done by gravity

We can then put that information into the work-energy theorem.

Total work done on an object = change in kinetic energy of the object

Work done by you – work done by gravity = change in kinetic energy of the rock

The change in kinetic energy of the rock is zero, so our relationship becomes

Work done by you – work done by gravity = 0

It's pretty easy to figure out how much work gravity does. The force of gravity on any object that has a mass m is equal to mg, where g is a special number that represents the acceleration of all objects near the surface of the Earth. If we raise the rock a distance h, then here's the work done by gravity on the rock:

Work done by gravity = (force of gravity) (distance moved in the direction of the force)
= $(mg)(h)$
= mgh

Look familiar? It should. Let's plug this back into the work-energy theorem.

Work done by you – work done by gravity = change in kinetic energy of the rock

Because the change in kinetic energy is zero, this becomes

Work done by you – mgh = 0

Adding mgh to both sides of this equation[5] gives us

Work done by you = mgh

This result should be a bit comforting. In a sense, you "gave" the rock gravitational potential energy by doing work on it, so the work you do on the rock really should equal mgh, our formula for the change in gravitational potential energy of an object.

I should let you know that in most science books the whole relationship between work and energy is done in reverse. The author defines what work is, and then shows that the work done on things gives expressions like mgh and $\frac{1}{2} mv^2$. What I did was try to justify (in Chapter 1) that mgh and $\frac{1}{2} mv^2$ were reasonable expressions for the thing called energy, and then showed how those expressions related to that odd definition of work.

[5] See the *Stop Faking It!* book on math.

One question that might be going through your head is, "What's the point?" Why have all these weird expressions for work and energy and why draw diagrams for how one changes into another? Well, you're about to find out. There's a very special property of energy, which I'll introduce in the next few sections, that makes the concept useful.

More science stuff–the sequel

I usually alternate the *things to do* sections and the *science stuff* sections, but I'm going to make an exception in this case. Before you do more activities, I want to tell a story that introduces that special property of energy I talked about and helps put all of this energy material into perspective. The story is adapted from one written by Richard Feynman in *The Feynman Lectures on Physics*.[6] Richard Feynman died a number of years ago. He was a remarkable guy who did a lot of remarkable things, including win a Nobel Prize. He was a physicist who was a professor first at Cornell and then Caltech, and he could explain physics concepts better than just about anyone, which explains why the *Feynman Lectures* still sells lots of copies. If you ever get really serious about learning physics, that three-volume set is the way to go. Of course, most of the material there is *way* beyond what we're talking about here but is still entertaining. Anyway, here's the story.

There once was an obsessive-compulsive mother (we'll call her Jane) who liked keeping track of her son's (we'll call him Elroy) toy blocks. She would count them every day, just to make sure they were all there. Every day, she counted 24 blocks. Things went along just fine until one day she counted only 21 blocks. This worried her no end, but after a while she noticed that Elroy's window was open. She looked out and, lo and behold, there were three blocks on the lawn. All was right with the world.

On another day, Jane noticed there were 28 blocks! Knowing that blocks don't reproduce, she finally figured out that Elroy's friend had been visiting, and had left four blocks behind. So Jane made up a rule: If the number of blocks is to stay at 24, no blocks will be allowed to enter or leave the house.

Time passed in block-secluded bliss until a day when Jane found only 20 blocks. She started to look in Elroy's toy box, but he threw a fit, saying he needed his privacy and that she couldn't look in his toy box. Jane knew there were blocks in the toy box so she figured out how to count those without looking inside. She knew the weight of the toy box without any blocks inside and she knew the weight of one block, so she devised the following formula:

---

[6] Feynman, R.P.; Leighton, R.B.; and Sands, M. 1964. *The Feynman Lectures on Physics, Volume 1.* Reading, MA: Addison-Wesley.

$$\text{Number of blocks in toy box} = \frac{\text{weight of toy box + blocks} - \text{weight of box without blocks}}{\text{weight of one block}}$$

Makes sense, because that difference in weight between a box with blocks and a box without blocks is the total weight of the blocks inside. Divide by the weight of a single block, and you get the number of blocks. So all she had to do was weigh the toy box, plug into this formula, and out would pop the number of blocks inside. Great!

Another day, after counting visible blocks and weighing the toy box, only 16 blocks were accounted for. Jane noticed, however, that the level of Elroy's bath water had risen since the previous day. One of Jane and Elroy's quirks was that they never changed Elroy's bath water. As you can guess, it was rather dirty after a couple of weeks—so dirty that you couldn't see anything submerged in it. Suspecting some blocks were in the bath water, Jane figured out how much the water level rose for each block submerged in the tub. She did this by adding one block and measuring the resulting change in water level. Then she used the following formula:

$$\text{Number of blocks in the tub} = \frac{\text{total rise in water level}}{\text{rise in water level per block}}$$

Everything was hunky dory, until one day when Jane couldn't find a single block. She weighed the toy box, measured the water level in the bath, and still couldn't account for all the blocks. Later that day, she noticed that all the toilets in the house were stopped up. She called the plumber, who removed the source of the problem (blocks!) and then charged her $5 for every block stuck in the system. This gave Jane an idea. She could figure out how many blocks Elroy stuffed down the toilets by using the following formula:

$$\text{Number of blocks in the plumbing} = \frac{\text{amount of plumber's invoice}}{\text{\$5 per block}}$$

Weeks later (same bath water!), Jane had a friend over for coffee. In the middle of conversation, Jane said, "Wait just a minute. I have to count Elroy's blocks." First Jane checked to see that no windows were open and that Elroy didn't have a friend over. Then she weighed Elroy's toy box and measured the water level in the bathtub. After calling the plumber to the house and paying him, Jane put the appropriate numbers into the following:

$$\text{Total number of blocks} = \frac{\text{weight of toy box} - \text{weight of box without blocks}}{\text{weight of one block}} + \frac{\text{total rise in water level in bath}}{\text{rise in level per block}} + \frac{\text{amount of plumber invoice}}{\text{\$5 per block}}$$

Putting down her calculator, Jane gave a sigh of relief and said, "Yep. 24 blocks!" Jane's friend decided that Jane had finally gone off the deep end because *there were no visible blocks!*

Jane might be certifiably loony but she's given us a great analogy for energy. You see, the blocks represent energy. Energy itself isn't visible and you can't hold "an energy" in your hand. But we can calculate all sorts of quantities, such as $\frac{1}{2}mv^2$, and *keep track* of energy. When we do keep track of the energy associated with a given system, we find that we always get the same total number, just as Jane always got the number 24 in "counting" Elroy's blocks.

One more point before I give you the big picture with energy. Remember that in order to get the same number every day, Jane had to make sure that no blocks left or entered the house. With energy, instead of defining the limits of a house, we define a **system**. You can think of a system as an imaginary boundary around the objects you're studying. So anyway, here's that remarkable principle I told you about.

In a closed system (no energy in or out), the total amount of energy remains constant.

The name for this principle is **conservation of energy**. I have no idea why that's the name. It has nothing to do with *conserving energy,* as in using as little energy as possible, and it certainly has nothing to do with being a conservative as opposed to a liberal. Whatever the source of the name, it means that for a given system (a collection of objects), every time you calculate the total amount of energy in the system, you get the same number.

Let's look at a couple of examples. Think about the pendulum you built as a closed system. For the first few swings at least, the pendulum keeps coming back up to the same initial height.

Figure 2.17

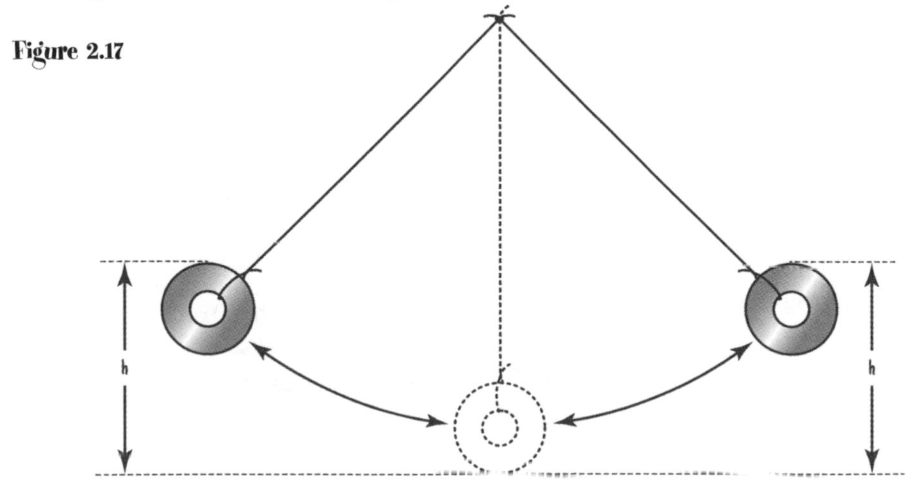

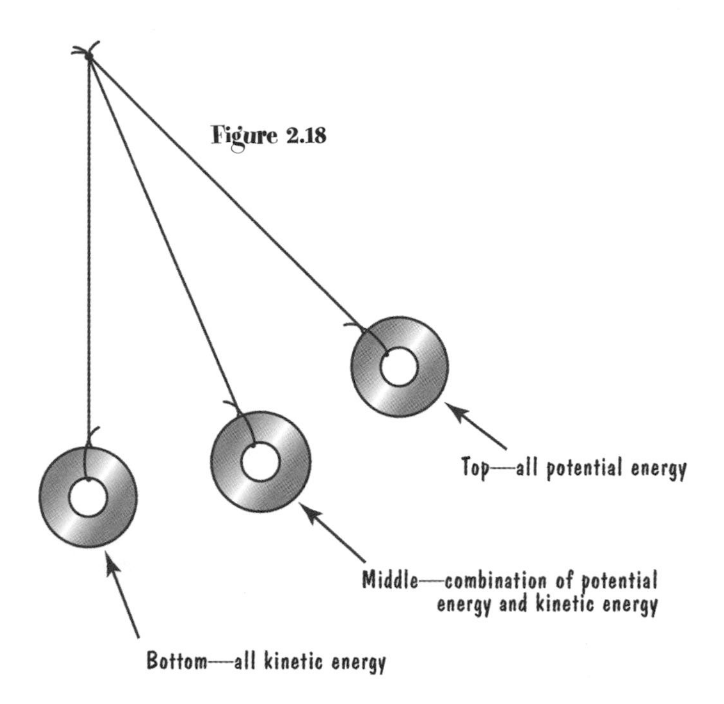

The energy of this system changes back and forth between potential and kinetic energy, but the total number you get from each quantity, *mgh* (gravitational potential energy) and $\frac{1}{2}mv^2$ (kinetic energy), will always give you the same number. That holds true at any point of the motion. What that means, in terms of the diagram, is that:

Potential energy at the top = potential energy + kinetic energy at the midpoint

= kinetic energy at the bottom

Using our formulas, this becomes:

$$(mgh)_{top} = (mgh + \tfrac{1}{2}mv^2)_{middle} = (\tfrac{1}{2}mv^2)_{bottom}$$

The *h* values and *v* values will be different at each point, but whatever they are, the number for the total energy should always have the same value.

I'm lying just a bit here. This isn't really a closed system because the pendulum loses energy to the surrounding air as the friction heats up the air. If we could calculate the rate at which the pendulum transfers energy to the air molecules, we could always have a constant number for the total energy.

On to "foreshadowing for the science geek." As you might recall, in the *Applications* section of the previous chapter, I outlined the following situation: When a rock falls off a cliff of a certain height, you can calculate the gravitational potential energy at the top of the cliff and the kinetic energy just before the rock hits the ground and get the same number for both kinds of energy.

That result should make all sorts of sense by now. All that happens is that the gravitational potential energy at the top gradually transforms into kinetic energy of the rock as it falls. Energy is conserved in a closed system, so the two numbers should be equal. Of course, all that falls apart when the rock hits the ground. Then all that kinetic energy of the rock transforms into thermal energy that leaves our closed system.

Even more things to do before you read even more science stuff

Find a few sections of Hot Wheels track, a marble or small ball, and a friend to help you. With the help of your friend, hold the connected sections so they form a loop, as in Figure 2.19.

Predict how high up the long side of the track you have to release the marble so it stays on the track all the way around the loop. Should you release it at the same height as the top of the loop? (See Figure 2.20 and remember conservation of energy.)

Figure 2.19

Or maybe you should place it higher up? Whatever you predict, go ahead and test your prediction. If your predicted release height doesn't work, experiment a while until you get the correct height.

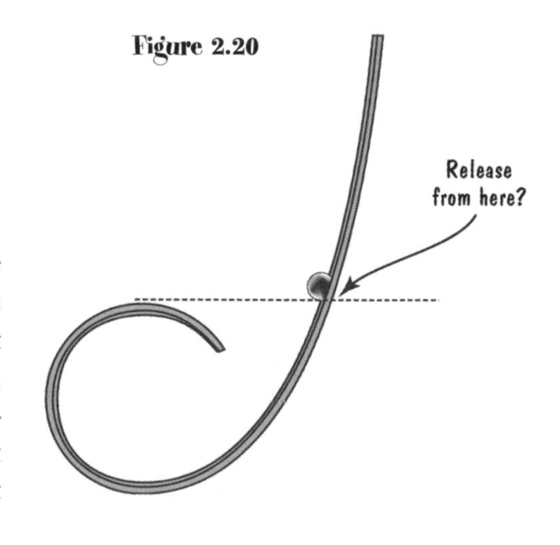

Figure 2.20

Even more science stuff

First, a question. Aren't you glad the people who design roller coasters that loop like this know their physics? Chances are you didn't release the marble nearly high enough on your first try. Why is that? After all, if conservation of energy really works, shouldn't the marble, like your pendulum, get right back up to the height it started?

Conservation of energy *does* work in this case, but you have to be careful that you include all the energy involved. To do that, let's call the release point position 1 and the top of the loop position 2 (Figure 2.21). Conservation of energy tells us that if we calculate the energy at these two different positions, we should get the same number. Of course that's not exactly true, because our system will lose thermal energy to friction, but we'll just ignore that for now.

Total energy of system at position 1 = total energy of system at position 2

Figure 2.21

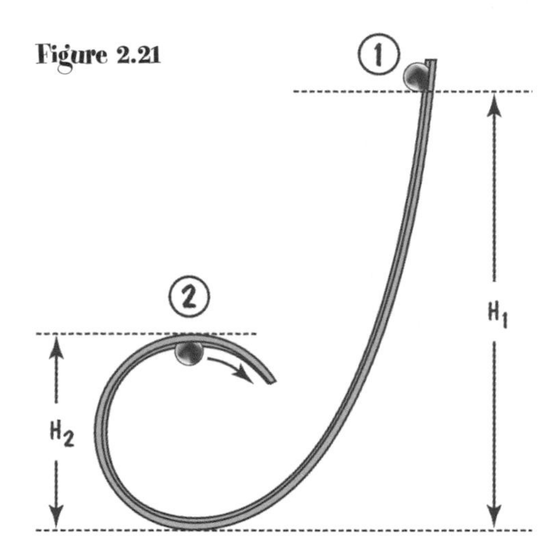

At position 1, the release point, it's easy to keep track of the energy. Nothing is moving, so all the energy is gravitational potential energy. At the top of the loop, it won't do to have all potential energy, because then the marble would have zero kinetic energy with a corresponding zero velocity. With zero velocity along the track, the marble would drop straight down off the track. We want the marble to keep going around the loop, so it has to have a velocity at that point (Figure 2.22).

Notice also that the marble *rolls* along the track instead of just sliding. That means that, in addition to moving with the velocity shown in the previous diagram, the marble has to be rotating. When you have an object that's rotating as well as moving along at a velocity *v*, it's convenient to separate its kinetic energy into two parts—the kinetic energy it has because of its velocity *v*, and the kinetic energy it has because of its rotation. We call these two kinds of energy *translational kinetic energy* and *rotational kinetic energy*.

Figure 2.22

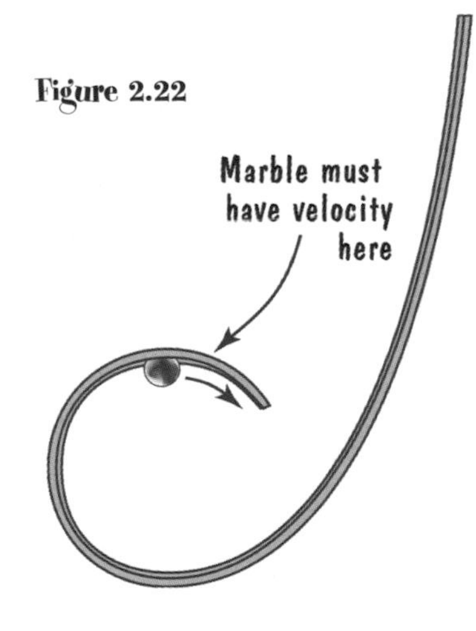

Now we can set up the energy relationship. To save space, I'm going to use *PE* to represent gravitational potential energy and *KE* to represent kinetic energy. In addition, KE_{trans} will represent translational kinetic energy (the kinetic energy the marble has because it's moving at a velocity *v*) and KE_{rot} will represent rotational kinetic energy.

PE at position 1 = *PE* at position 2 + KE_{trans} at position 2 + KE_{rot} at position 2

Before going on, notice that this relationship tells us why you had to release the marble from so high up the long side of the track. The potential energy it had there transforms not only into potential energy at the top of the loop but also into translational and rotational kinetic energy at the top of the loop.

I'm going to take the conservation of energy relationship one step further and plug in the formulas we have for potential and kinetic energy. H_1 and H_2 will represent the height of positions 1 and 2 above the ground, v_2 will represent the velocity of the marble at position 2, and m represents the mass of the marble. I'm also going to use the expression $^1/_5mv_2^2$ for the rotational kinetic energy at position 2. You'll just have to accept that this is the correct formula, because it would take *way* too much of our time to get into how I figured it out. So anyway, here's the conservation of energy relationship.

$$PE \text{ at position 1} = PE \text{ at position 2} + KE_{trans} \text{ at position 2} + KE_{rot} \text{ at position 2}$$

$$mgH_1 = mgH_2 + {}^1/_2\, mv_2^2 + {}^1/_5 mv_2^2$$

If you're about to scream because that looks really complicated, hold on just a bit longer. I'm going to do an algebra step that you might or might not remember from whenever you took that subject. It's called *canceling*. I can divide each side of that equation above by the mass, m. When I do that, all the m's cancel out of the equation, and we're left with:

$$gH_1 = gH_2 + {}^1/_2 v_2^2 + {}^1/_5 v_2^2$$

It doesn't really matter whether you followed what I just did. What matters is the result. And the result is this: When you write out the conservation of energy relationship for a marble rolling on a track, that relationship has nothing to do with how big the marble is or what its mass is. In other words, if you have 10 marbles, all of different sizes and weights, you would release all of them at the same point on the track in order for them to stay on the track as they go around the loop.

And your next question—"So what?" Well, think about how this applies to roller coasters. If a coaster has an upside down loop in it, then the cars have to have a certain amount of energy in order to negotiate the loop successfully. It would be a nightmare to design such a coaster if the necessary energy amount depended on the total mass of people in the coaster. You would either have to make sure you always had the same total mass of passengers or you'd have to adjust the height and speed of the coaster depending on how many people were riding and how heavy they were. Fortunately, it's just not a problem. The specifications are the same no matter what kind of passenger load you have. Of course, in a real coaster you'd have to be more careful with your calculations. I ignored friction in mine, and that's not something you can do in the real world!

One final thing. I mentioned in Chapter 1 that the h in mgh is due to an arbitrary choice of a reference level. For our loop problem, I took H_1 and H_2 to be the heights as measured from the bottom of the track. In other words, I chose the bottom of the track to be my *reference level*—the place where h would equal zero and an object would have zero gravitational potential energy.

I could have chosen the top of the loop as the reference level, and everything would have been okay. The reason any choice of reference level is okay is that, as I mentioned in Chapter 1, only *changes* in gravitational potential energy are important. As long as all the h's are measured from the same point, there isn't a problem.

Chapter Summary

- Energy readily transforms from one form to another.
- Work is defined as the force applied to an object, multiplied by the distance the object moves in the direction of the force.
- When you do work on an object, you change its energy. The work done is numerically equal to the change in energy.
- The total energy in a closed system (no energy in or out) is constant. This is known as the principle of conservation of energy. When you use formulas for different kinds of energy, you can numerically keep track of the energy in a system and use those results to predict things such as the velocity at which something will move at a later time or the height it will reach at a later time.

Applications

1. Stand in one spot and hold a full gallon jug of milk or water out in front of you. Do this for very long and you'll definitely be working! But you're not doing any work according to our definition of work because the jug isn't going anywhere, so the distance moved is zero. What's up with that? The key here is that you are not doing any work *on the jug*. Inside your body, all sorts of work is happening. Chemical reactions are taking place and muscles are contracting. Chemical potential energy is transforming to all sorts of other kinds of energy.

Figure 2.23

"If this is skim, what does whole milk weigh?"

2. Let's take a more general look at roller coasters as a transformation of energy. In case you don't get out much, I'll let you know how coasters work. It's pretty simple. There's a big

chain that takes the coaster to the top of the first hill. This is almost always the highest point on the coaster by far. That's because the gravitational potential energy the coaster gets by being at the top of the first hill is all the energy it will ever have.

From that point on, the coaster just coasts (hence the name!) along the track until the finish. It's a continual transformation between potential and kinetic energy, with some of the energy being in the rotational kinetic energy of the wheels of the coaster. Of course, any friction that's present helps transform some of the energy into thermal energy throughout the course. If the coaster is going to make it to the finish, you have to make sure the initial potential energy is enough to take into account all those transformations.

Figure 2.24

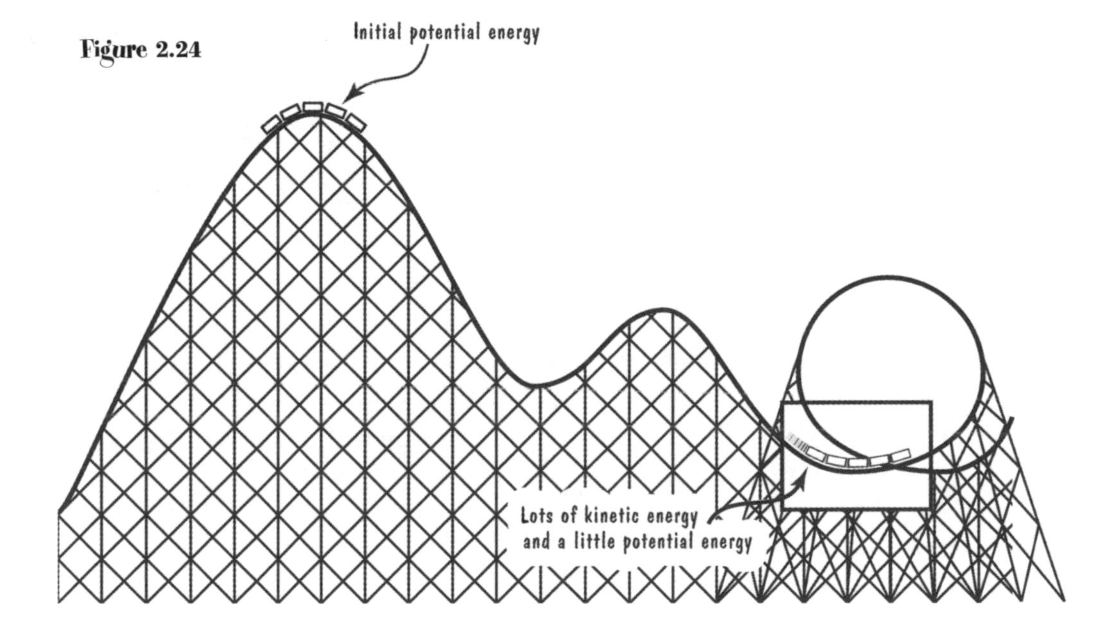

3. You can use conservation of energy to figure out how fast a rocket needs to be traveling in order to reach a certain orbit or to escape the pull of the Earth's gravity altogether. You have to use a more general form of gravitational potential energy than *mgh*. It's one that takes into account the gravitational force between the rocket and the Earth no matter how far apart they are. You also have to know a few things about the speed necessary for an object to remain in orbit at a given height above the Earth's surface. Once you figure that out, it's just a matter of setting the rocket's total energy after it takes off equal to the rocket's total energy when it's in orbit. And no, I won't go through that calculation. You're welcome.

4. What the heck, let's do one calculation. You throw a ball straight up with a velocity of 10 meters/second. How high above your hand will the ball go before falling back down (Figure 2.25)? We're interested in two different positions. We know the velocity of the ball as it leaves your hand and we want to know the height at the point where the ball stops and starts to fall back down. What I'm going to do is figure out the energy at each position and set those equal to each other.

Energy at position 1 = energy at position 2

First, I'm going to *choose* the zero point of gravitational potential energy (the reference level) to be your hand. This means that, just after the ball leaves your hand, it has kinetic energy but no gravitational potential energy. At the top (position 2), the ball has gravitational potential energy but no kinetic energy (it's not moving). (See Figure 2.26.)

$$(\text{Energy})_{\text{position 1}} = (\text{energy})_{\text{position 2}}$$

$${}^{1}/{}_{2}mv_1^2 = mgh_2$$

Figure 2.25 **Figure 2.26**

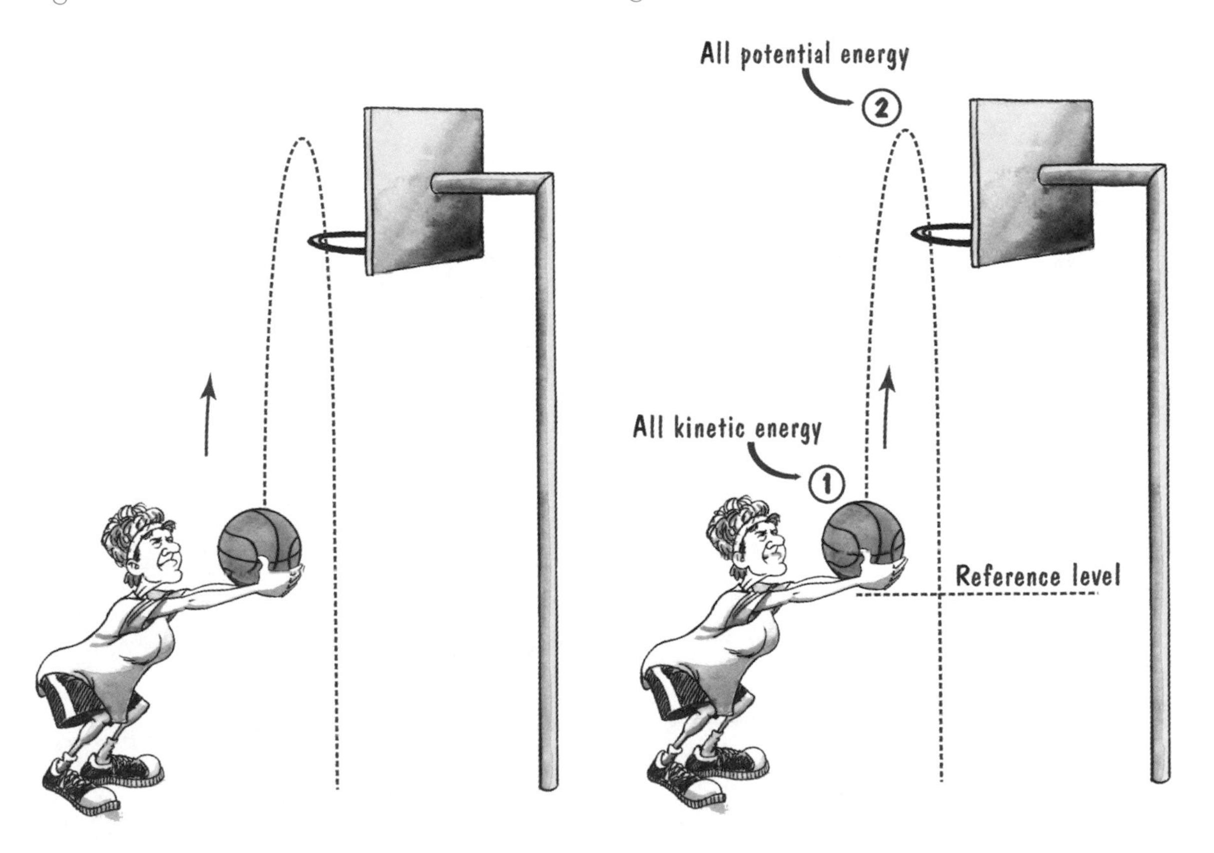

If algebra really freaks you out, either skip to the solution or close your eyes and flip to the next chapter. For all you other brave souls, I'm going to solve this equation for h_2. First I'm going to divide both sides of the equation by the mass of the ball, *m*.

$$\frac{\frac{1}{2}mv_1^2}{m} = \frac{mgh_2}{m}$$

The *m*'s on each side cancel, and we get:

$$\frac{1}{2}v_1^2 = gh_2$$

Now I'm going to divide both sides by *g*.

$$\frac{\frac{1}{2}v_1^2}{g} = \frac{gh_2}{g}$$

Now the *g*'s cancel on the right, and we're left with:

$$\frac{\frac{1}{2}v_1^2}{g} = h_2$$

So that's it: h_2 is equal to that expression on the left. All that's left is to put numbers in for v_1 and *g*.

$$h_2 = \frac{\frac{1}{2}(10)^2}{(9.8)}$$

$$= 5.1 \text{ meters}$$

It Slices, It Dices–It Gathers Dust!

In case you can't tell from the title, this chapter is about machines. No, not salad shooters and makers of julienne fries (although we *could* discuss those) but rather everyday things such as scissors and bottle openers and car jacks. These are known as **simple machines**, and they involve straightforward applications of work, energy, and the conservation of energy.

Before starting, I'll let you in on a little secret—I always *hated* learning about simple machines. Inclined planes, fulcrums, loads, mechanical advantage—boorrrrring! But then I found out that, like so many other science subjects, it wasn't the machines that were boring. It was the presentation of the subject in

"Step right up! See the Amazing Simple Machines! Be astounded as they perform mazimum feats of work with minimal amounts of energy!"

textbooks. Too much memorizing of formulas and not enough fun things to do. I'll try to stick to the fun activities in this chapter.

Things to do before you read the science stuff

Gather together a ruler (wood or hard plastic, not flexible), a pencil, and something small and sort of heavy, such as a paperweight, a rock, or one of Grandma's biscuits. Because "rock" is the shortest word to type, I'll assume from here on that you're using a rock. First just lift the rock to get a feel for how big a force you need to exert to lift it. Then place the pencil on a flat surface with the ruler on top so the pencil is at the center of the ruler. Put the rock on top of the ruler at one end and you should have the situation shown in Figure 3.1.

Figure 3.1

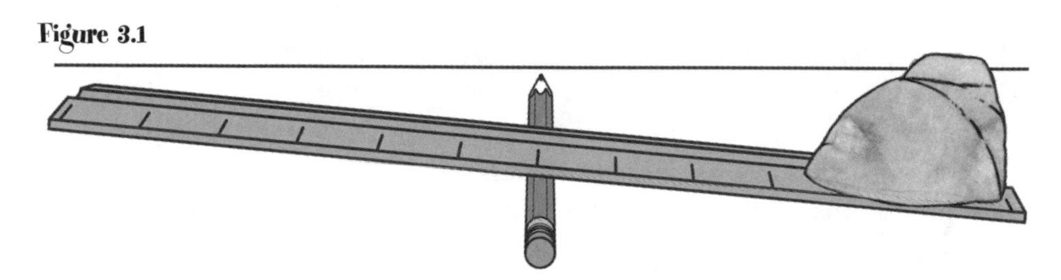

Push down on the free end of the ruler so the rock rises up. If you didn't listen to me and used a flexible ruler, now you know why you need a stiff ruler.

Figure 3.2

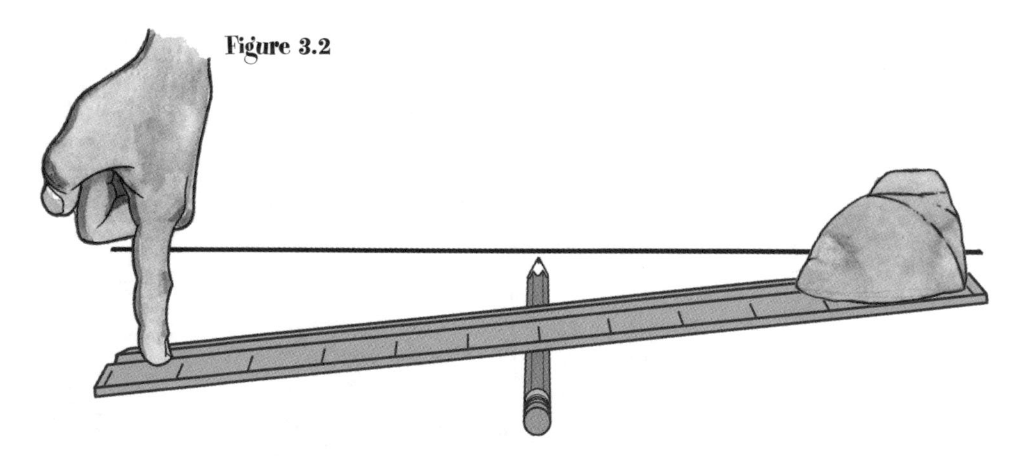

Compare how hard you have to push down with how hard you had to pull in order to lift the rock directly. Also compare the *distance* you push the ruler down with the distance the rock rises. No need to measure; just eyeball it (Figure 3.3).

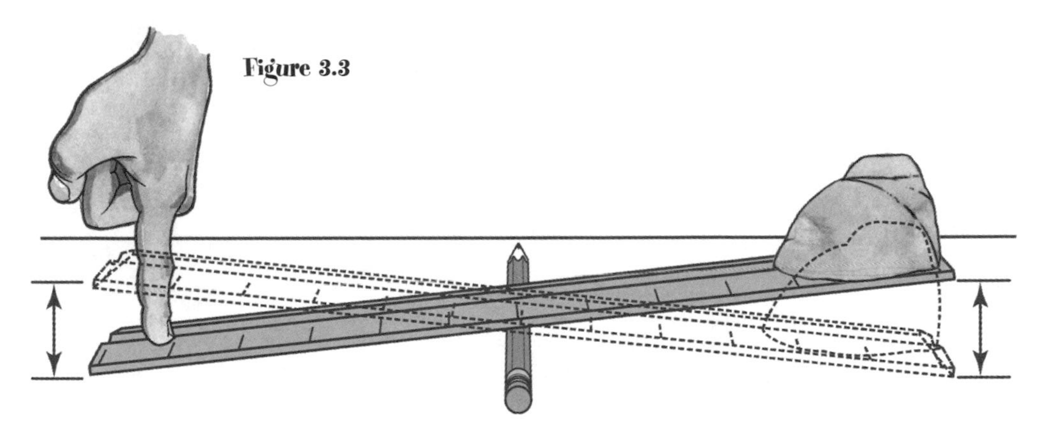
Figure 3.3

Reposition the ruler and pencil as shown in Figure 3.4 and repeat all the previous steps—push down on the ruler, lifting the rock; compare the push with the earlier one; and compare the distances.

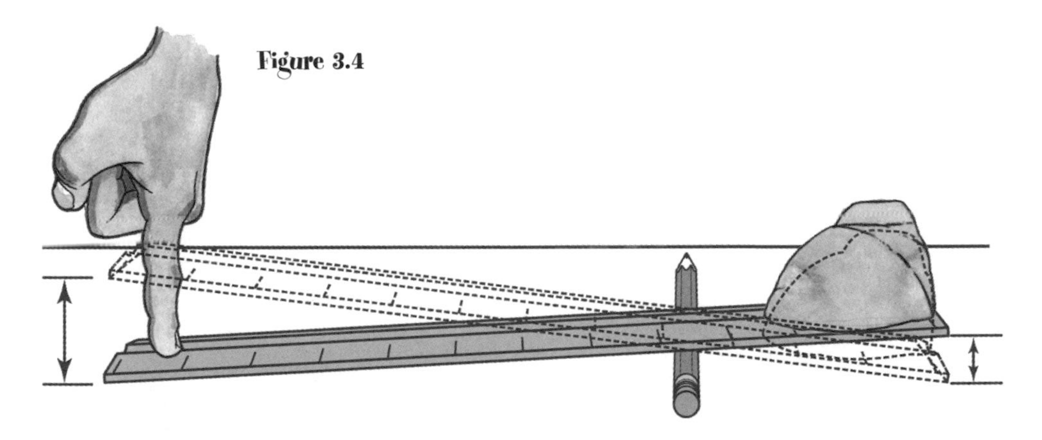
Figure 3.4

Now reposition everything as in Figure 3.5 and repeat the same steps.

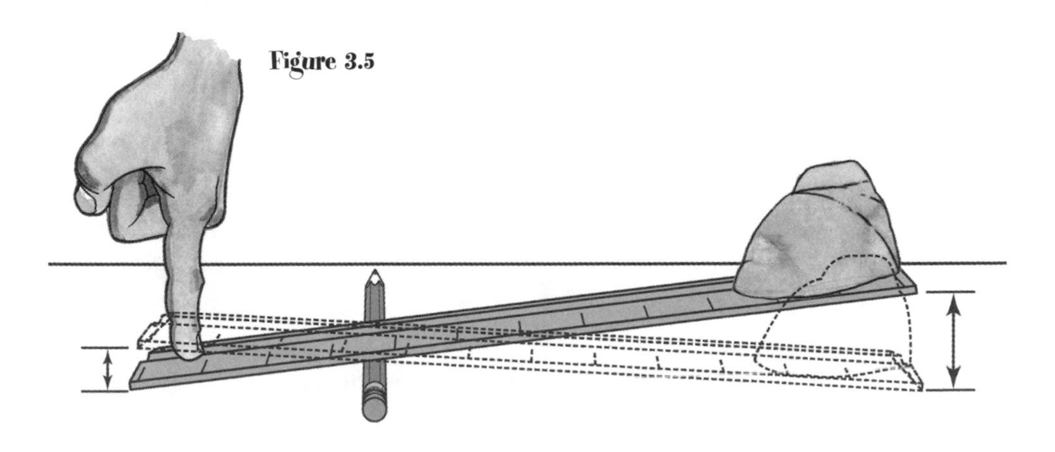
Figure 3.5

Craft time. Get a pair of scissors and a sheet of heavy stock paper, such as an index card. Try cutting the card using the scissors in the two ways illustrated in Figure 3.6.

Figure 3.6

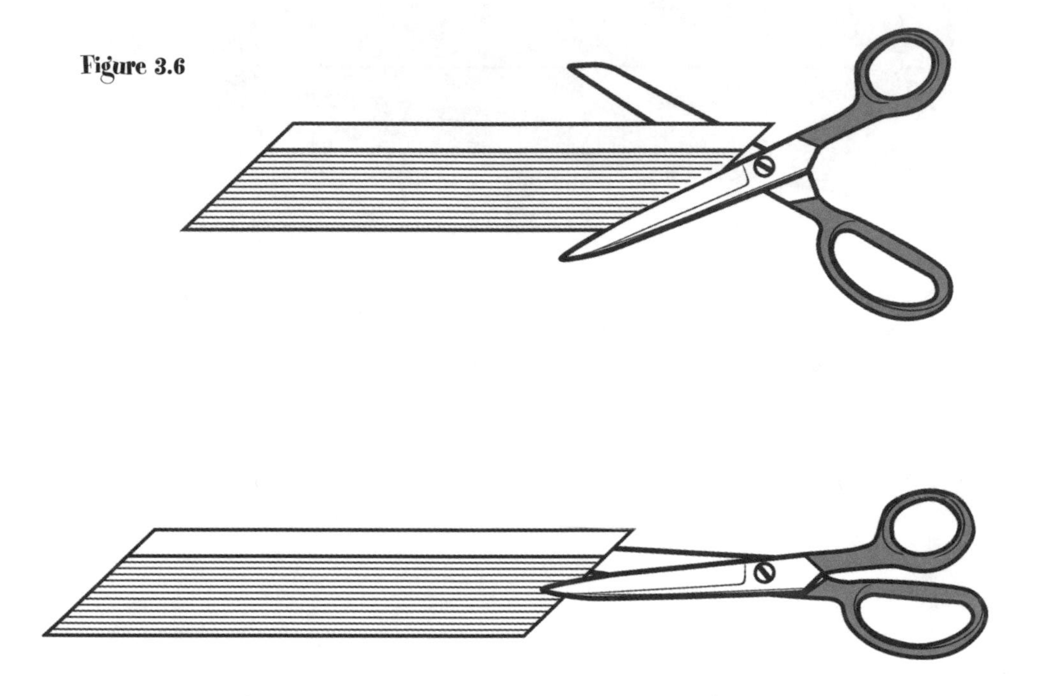

Which method is easier for cutting the card?

The science stuff

Let's start with the obvious. Each time you used the ruler to lift the rock, the ruler changed the direction of the force you exerted. You pushed down, and the rock went up. Magic, huh? No, not magic, but it is one thing simple machines are good for–they can change the direction of the force you apply.

You undoubtedly noticed that in some situations it was easier to lift the rock than in other situations. When the pencil is near the rock end of the ruler, it's much easier than when the pencil is near the end of the ruler you're pushing on. When the pencil is in the middle of the ruler, you have to push about as hard as you would have to in order to lift the rock directly (that might or might not have been obvious to you).

Work and energy to the rescue to explain what's going on. In each case, you changed the rock's gravitational potential energy by raising it up. You did work on the ruler because you exerted a force on the ruler, and the end of the ruler you were pushing moved in the direction of that force. In turn, the opposite end of the ruler did work on the rock.

Figure 3.7

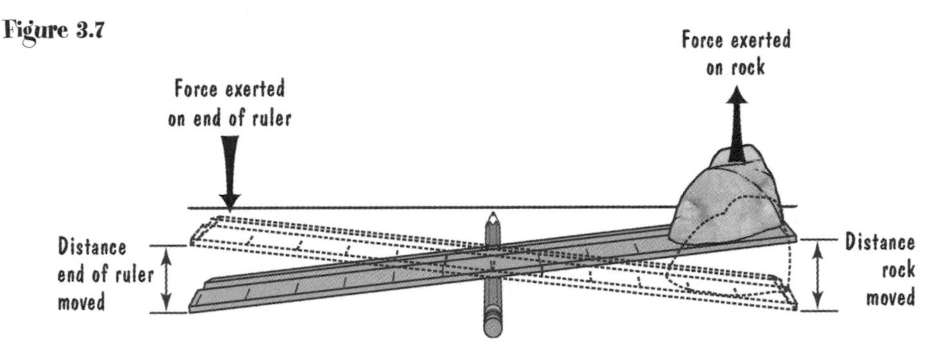

Think of the pencil, ruler, and rock as a closed system. You did work on this system, giving it energy. Let's track that energy. A bunch of it obviously showed up as an increase in the gravitational potential energy of the rock. Friction between the ruler and pencil sent some energy to thermal energy, and the ruler moved so a tiny bit went into kinetic energy of the ruler.

Figure 3.8

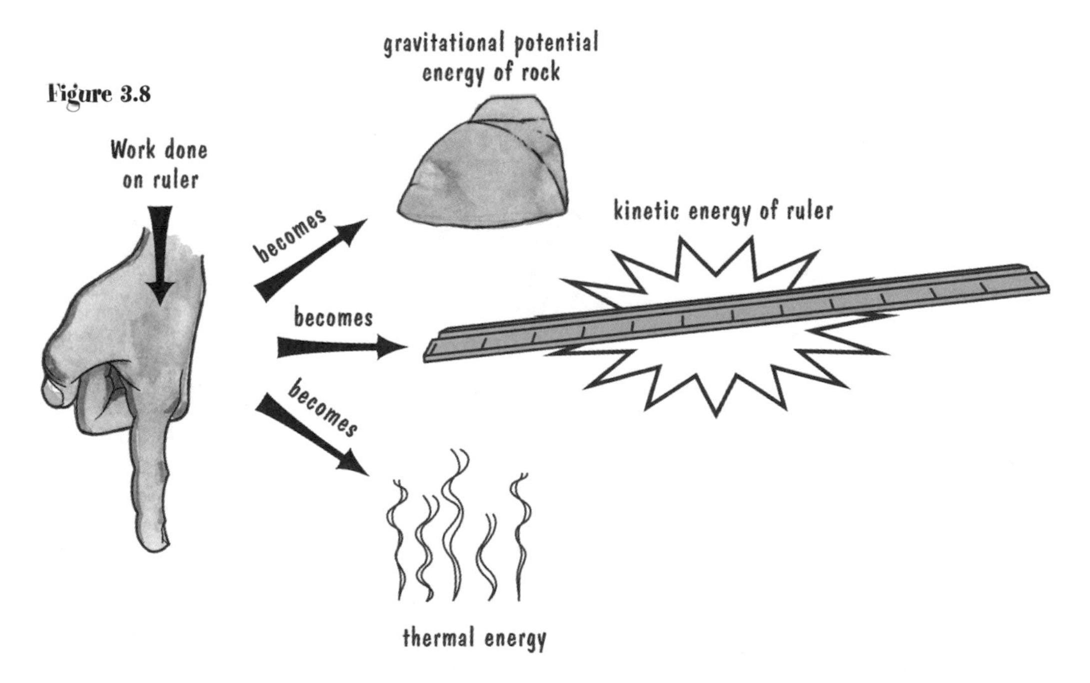

You're just going to have to trust me on this next fact. The vast majority of the work done on the ruler transformed into a change in gravitational potential energy of the rock. We also know that the work the ruler did on the rock equals the change in gravitational potential energy of the rock. And that means that the work you did on the ruler is equal to the work the ruler did on the rock.

Work done on the ruler = work done on the rock

We could also write this as:

Work done on the system = work done by the system

Or, put another way, this means:

Work in = work out

Keep in mind that this isn't 100 percent true, due to the energy "losses" outlined in Figure 3.8. For now, though, we'll assume it's exactly true. In terms of our formula for work, this is:

$$F_1d_1 = F_2d_2$$ [1]

where F_1 equals the force exerted by you on the ruler, F_2 equals the force exerted by the ruler on the rock, d_1 equals the distance your end of the ruler moves, and d_2 equals the distance the rock moves. Notice that the distance moved in each case is in the direction of the applied force, so we don't have to worry about that "parallel" thing.

Figure 3.9

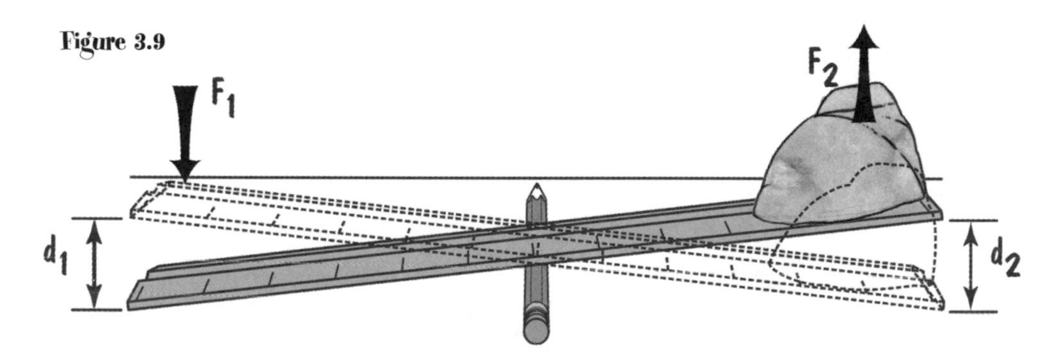

Let's look at the three situations you investigated with the pencil, ruler, and rock.

1. Pencil in the middle of the ruler. In this case, the distance the rock moves is equal to the distance you moved the opposite end of the ruler. Don't believe me? Try it again and measure. This means that $d_1 = d_2$.

   What does that imply for the forces F_1 and F_2? Well, if $F_1d_1 = F_2d_2$, and $d_1 = d_2$, then F_1 *must* equal F_2. If you don't see that, just plug in some numbers.

---

[1] I should let you know I'm making an approximation here. If you watch carefully, the end of the ruler doesn't go straight down. It goes in an arc, because the ruler is actually rotating. If we analyze this as a rotation, though, I have to introduce angular measures and something called torque. That would take at least an extra chapter, so I'm not going to do it. Suffice to say that, as long as d_1 and d_2 aren't very large, the approximation is a good one.

Suppose d_1 and d_2 are both 3 centimeters (remember, they're equal!). Then the work relationship says:

$$(F_1)(3) = (F_2)(3)$$

There is no way you can put in different numbers for F_1 and F_2 and end up with a true statement. Try it with $F_1 = 2$ and $F_2 = 4$. Does 6 equal 12? Nope. So equal distances means equal forces.

2. Pencil closest to the rock. In this case, d_1 is quite a bit larger than d_2.

Figure 3.10

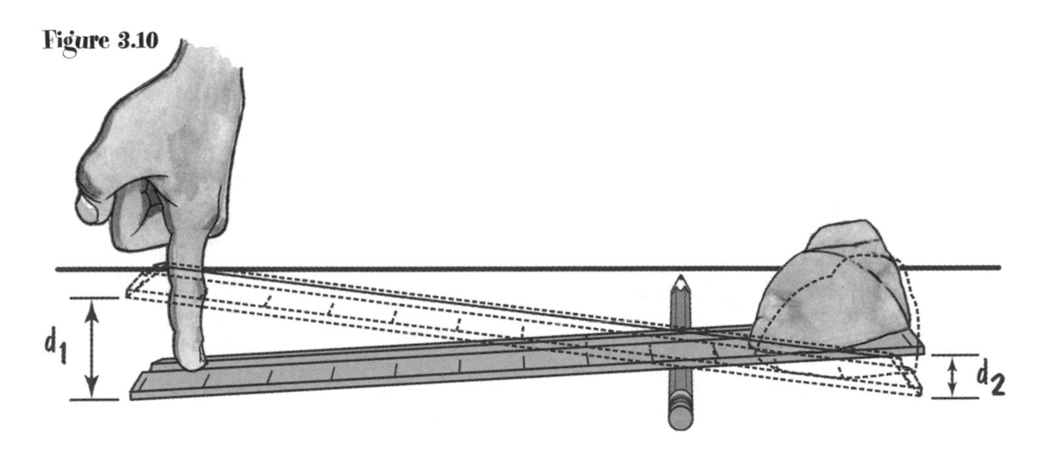

What does that say about F_1 and F_2? To find out, we use the work relationship again, which says that work in equals work out.

$$F_1 d_1 = F_2 d_2$$

We need d_1 to be larger than d_2. I'll just choose arbitrary numbers and let d_1 equal 4 centimeters and d_2 equal 1 centimeter. Then we get:

$$(F_1)(4) = (F_2)(1)$$

Further, suppose it takes a force of magnitude 8 to lift the rock. Then F_2 equals 8, and we have:

$$(F_1)(4) = (8)(1)$$

To make this equality work, F_1 has to be equal to 2.

$$(2)(4) = (8)(1)$$

So you are able to exert a force of 8 on the rock by exerting only a force of 2 on the ruler. Cool, you increased your strength! This increase in force happens no matter what values the F's and d's have. If d_1 is greater than d_2, then F_2 is greater than F_1. Try a few other numbers to see that this is true.

You should have noticed this change in force. With the pencil near the rock, it's pretty darned easy to lift the rock by pushing down on the ruler.

Unfortunately, though, you don't get something for nothing. In order to get that increase in force, there is a trade-off in the distance. You had to push the ruler down much farther than the rock went up. A trade-off between force and distance is the basis for the operation of all the machines we'll deal with in this chapter. You can get an increase in force, but it will cost you some distance.

3. Pencil near the end on which you push down.

Figure 3.11

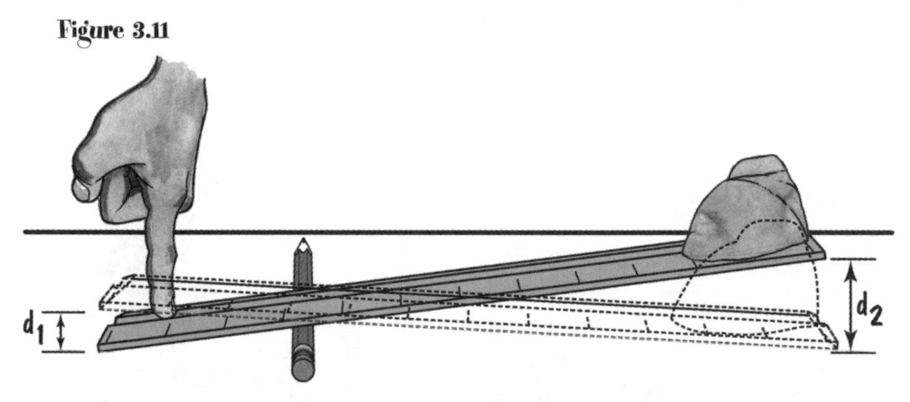

In this case, d_1 is *less than* d_2. Our work relationship, $F_1d_1 = F_2d_2$, tells us that F_1 must be *greater than* F_2. In other words, we get a decrease in force. You undoubtedly noticed that in this situation you had to push down much harder to lift the rock. That's because there's a different kind of trade-off. You pushed harder than you normally would to just lift the rock, but the rock changed its height more than the distance you pushed down. You pushed harder but got more distance for your effort.

For the record, the accepted name for this pencil and ruler machine is a **lever.** Let's apply what we now know about that lever to the operation of a pair of scissors. A pair of scissors is actually two levers that move in opposite directions.

Figure 3.12

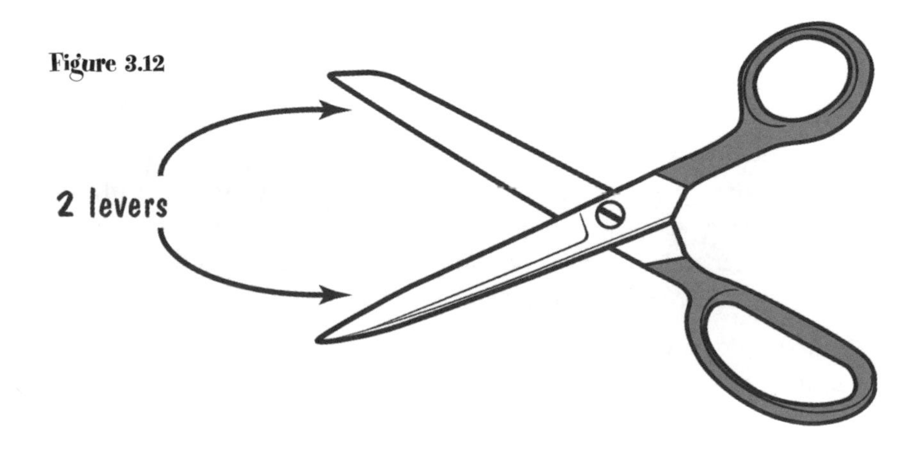

These happen to be sharpened levers, which is a good thing given their purpose. It's really difficult to cut things using two rulers. Anyway, the two levers rotate about the place where they're joined. This is the equivalent of the pencil in the previous example, and is known as the **fulcrum**.

When cutting paper (refer back to Figure 3.6), you want the force the two sides exert on the paper to be greater than the force you exert on the scissor handles. In order to accomplish that, you need to place the paper near the fulcrum of the scissors. Why? Because at that position, you move the handles much farther than the parts of the sharpened levers near the fulcrum move.

Figure 3.13

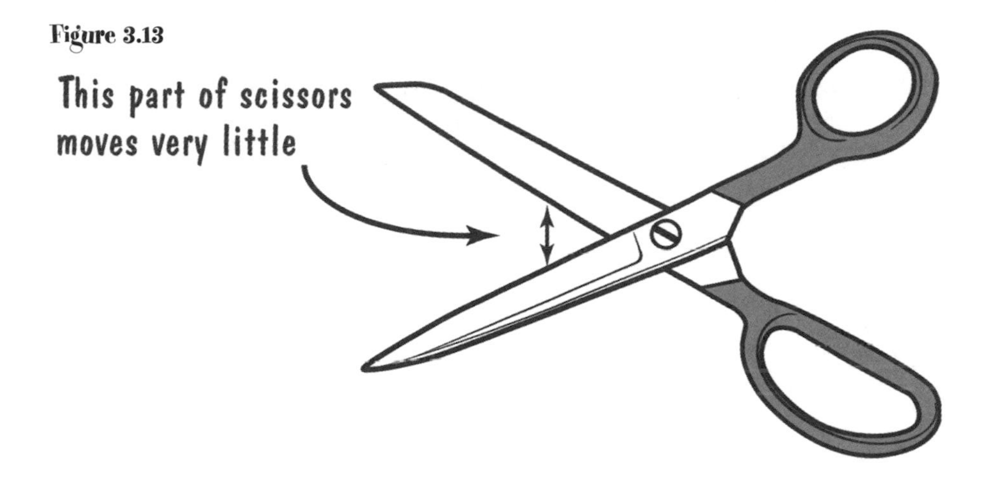

$F_1d_1 = F_2d_2$ then tells us that the force exerted by the sharp part of the scissors is greater than the force exerted by you on the handles. Move the paper to the tip of the scissors, and the scissors actually exert a smaller force than you exert on the handles. Makes it tough to cut heavy paper.

One final note about levers. In many elementary science textbooks, you will find something called the *law of the lever*. That law equates one force times a distance with another force times a distance. Although that might look identical to what we've done here, it's not. I analyzed the lever from an "energy perspective," while the *law of the lever* involves equating two things called *torques*. The distances involved in the *law of the lever* are not the same as the distances I've discussed here, even though the forces are the same.[2]

Before going on to the next section, here's a quick summary. Conservation of energy applies to simple machines such as a lever. By ignoring the

[2] The distances used in the law of the lever are the distances from the fulcrum to the points at which forces are applied. These distances often are referred to as *moment arms*.

energy losses in a machine (usually okay to do, but not always), we can say that:

Work you put into the machine = work done by the machine

or

Work in = work out

Because work is force times distance, this can be written as:

$$F_1d_1 = F_2d_2$$

Many simple machines involve a force/distance trade-off. You can trade distance for force, or vice versa.

More things to do before you read more science stuff

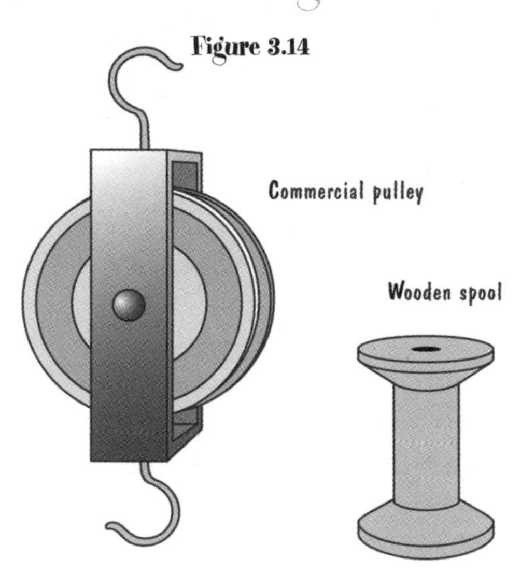

Figure 3.14

For this section, you need a couple of small, plastic pulleys. Commercial ones look like Figure 3.14.

If you don't have access to commercial pulleys, have no fear. Head to a hobby or craft store and get a couple of small wooden spools. Get the really tiny ones that are about 3 centimeters long. You'll also need a couple of metal paper clips.

Bend one of the paper clips as shown in Figure 3.15.

Figure 3.15

Place a spool over the horizontal part of the bent paper clip as shown in Figure 3.16.

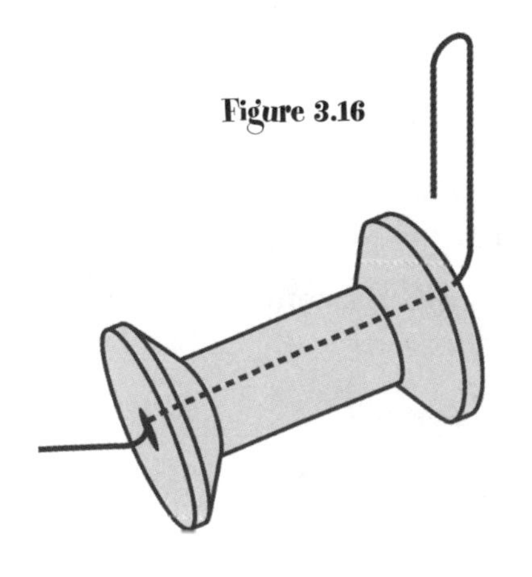
Figure 3.16

Now use a pair of pliers (a set of levers!) to bend the tip of the paper clip slightly, so as to hold the spool in place. Make sure the spool can spin freely on the paper clip.

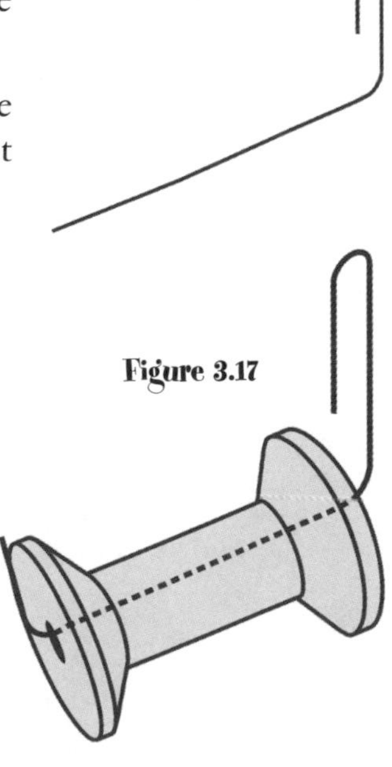
Figure 3.17

Voilà! A pulley. While you're at it, make a second pulley.

Figure 3.18

Now hang your commercial or homemade pulley from someplace sturdy. A drawer handle on a filing cabinet or something similar works well (Figure 3.18).

Figure 3.19

Grab one of those pendulums you made in Chapter 2. Hang the second pulley on the end of the pendulum.[3] Hold onto the end of the string and get a feel for how hard it is to lift the weight straight up (Figure 3.19).

Now thread the string over your pulley as shown in Figure 3.20. Using this arrangement, lift the weight. Compare how difficult this is with how difficult it is to lift the weight directly. Next, change your system so it looks like Figure 3.21.

Figure 3.20

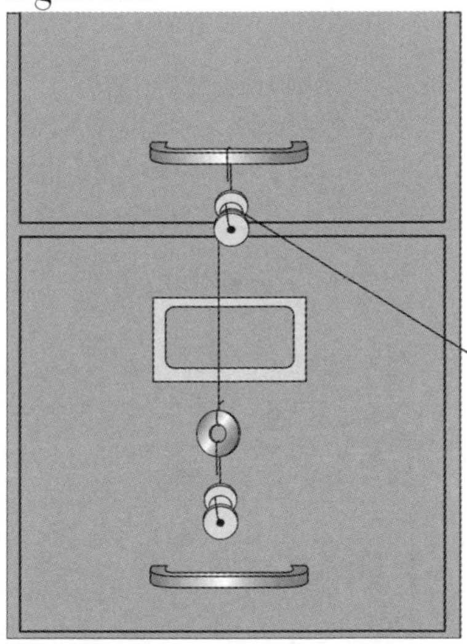

When you pull on the string, the combination of pulley and weight should move upward. How hard do you have to pull to lift the weight? Compare with how hard you had to pull using the setup in Figure 3.20. How far do you have to pull the string compared with how high the weight rises?

More science stuff

As you might have guessed, pulleys involve a trade-off between force and distance, just as with a lever. A single pulley won't do that for you, though. All you can do with a single pulley is change the direction of the force you exert. That can be useful, but why stop there? With two pulleys set up as in Figure 3.21, you clearly don't have to pull as hard as you do when using a single pulley. Of course, you have to pull the string about twice as far, so again, you don't

Figure 3.21

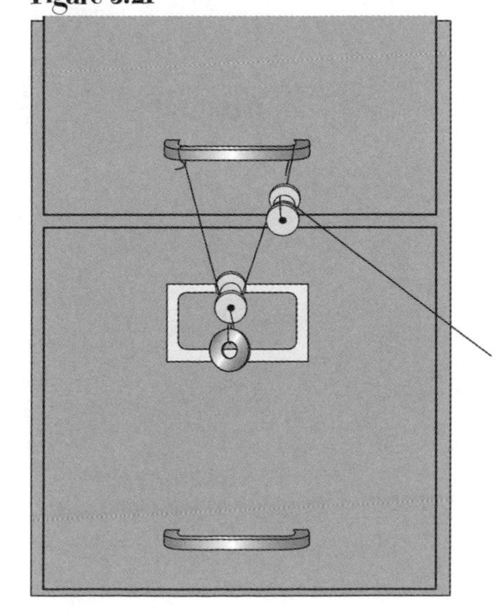

[3] This might seem a silly thing to do but it ensures you lift the same weight here as in a later situation. That's important for comparison purposes.

get something for nothing. The work relationship looks like the following:

Work in = work out

(Force you apply)(distance you pull string) = (force pulling on weight) (distance weight moves)

In the two-pulley setup, you pull the string about twice as far, and the force exerted on the weight is about twice as strong as your pull (Figure 3.22).

Now you just might be asking, "How in the world did this setup double the force I applied?" Glad you asked. With a single pulley, the string just more or less transmits to the weight the force that you apply (Figure 3.23).[4]

Using two pulleys, however, the weight is now supported by two sections of string (Figure 3.24). This effectively doubles the force you apply.[5]

Figure 3.22

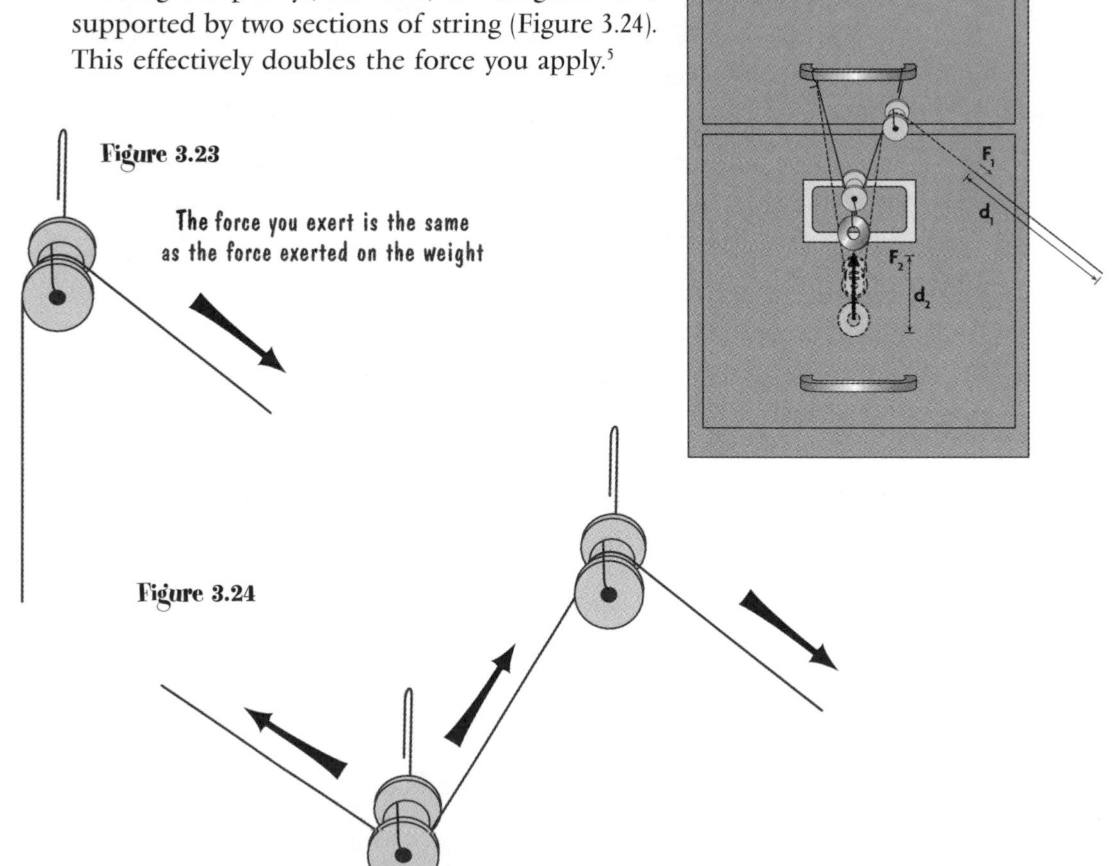

Figure 3.23

Figure 3.24

[4] There's a subtlety here that makes this statement untrue if the string and weight are accelerating (changing their speed). We won't get into that subtlety as it's way beyond the scope of this book.

[5] Because this setup isn't straight up and down but angled, you don't get a doubling of the force. Instead, you get something less than that.

Figure 3.25

Watch the supporting strings as you use this setup and you should also see why the weight doesn't rise as far as you pull the string.

You can get an even greater increase in force by adding a third pulley, as in Figure 3.25.

You can keep adding pulleys and increasing the force that's applied to the weight. Each time you do that, though, you increase the distance you have to pull compared with how high the weight rises (Figure 3.26).

When you combine lots of pulleys, you have what's known as a **block and tackle**. Auto mechanics use a block and tackle in order to lift an entire engine out of a car. Try that job without using a machine!

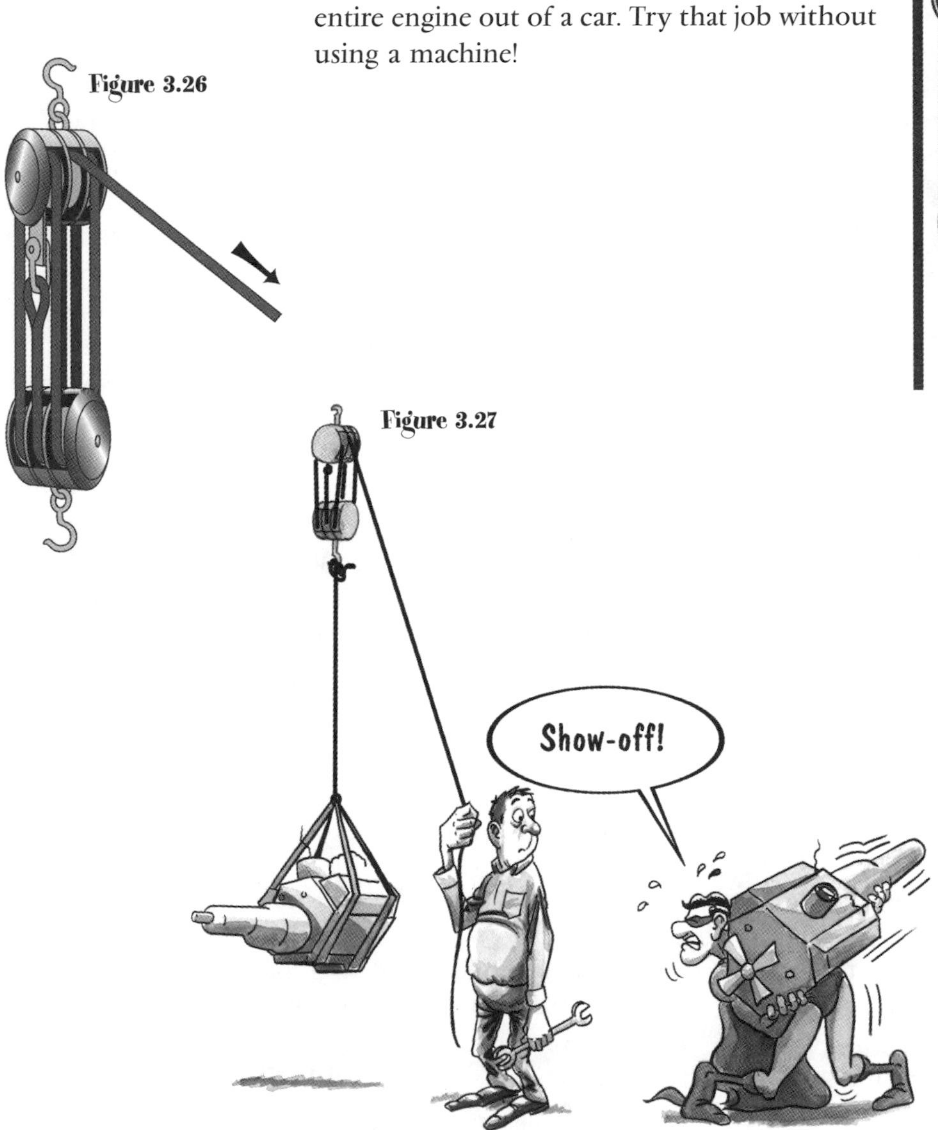

Figure 3.26

Figure 3.27

Even more things to do before you read even more science stuff

This section is easy. Just take a look at the "perpetual motion" machine shown below. Such machines are supposed to keep going and going and going, without any input of energy. See if you can figure out how this machine is supposed to work. Then try to use the principle of conservation of energy to explain why the machine *shouldn't* work.

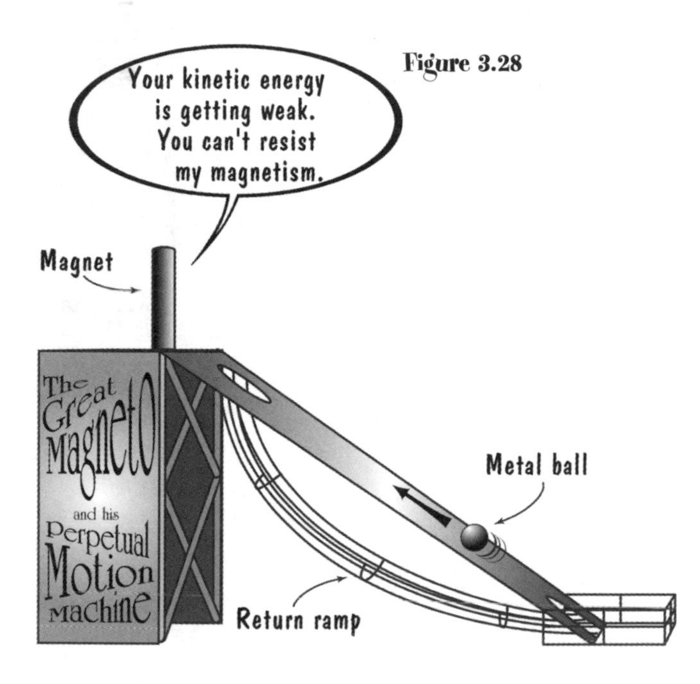

Figure 3.28

Even more science stuff

The basic idea I've used in this chapter is that work in = work out. But we know that isn't true! When you do work on a system, a chunk of that work shows up as work done by the system on something else, but certainly not all of it. There's always friction around, so that converts some of your input of work to thermal energy, which is eventually lost to the surrounding air. In addition, some of the work input shows up as kinetic energy of the parts of the machine, as in rotating pulleys.

"Work in" is *not* equal to "work out." In fact, the work out is always less than the work in because all machines have some kind of energy "losses." That leads us to the concept of **efficiency**. A really efficient machine is one in which there isn't a whole lot of wasted energy and the work out is *almost* equal to the work in. There's even a formula for efficiency and here it is:

$$\text{Efficiency} = \frac{\text{work out}}{\text{work in}} \times 100\%$$

Let's try to make sense of that formula. If the machine is 100% efficient (not really possible), then "work in" really does equal "work out." That means that (work out)/(work in) = 1. [Note: Anything divided by itself is equal to 1. If these two things are equal, then this fraction should equal 1.] Then we have:

$$\text{Efficiency} = 1 \times 100\%$$

which equals 100%. Now suppose half of the work you put into a machine shows up as wasted thermal energy. Then only half the work is left over for the work out, which means that (work out)/(work in) equals ½. Then we have:

$$\text{Efficiency} = \frac{1}{2} \times 100\%$$

which equals 50%, meaning this machine is 50% efficient.

How do you make an efficient machine? Easy, at least in principle. Reduce the friction as much as possible and have as few moving parts as possible.

All right, what about that perpetual motion machine? If it keeps on going without any input of energy, then it must be 100% efficient. Here's how it's supposed to work. The magnet draws the iron ball up the top ramp. When the ball reaches the hole at the top, it falls back down to the bottom via the lower, curved ramp. Then the magnet once again pulls the ball up the top ramp.

Very nifty and very unworkable. I'm thinking this will work one time. The ball is attracted to the top of the upper ramp and reaches the hole. When the ball reaches the hole, it might very well fall down to the lower ramp. But will it make it all the way to the bottom so it can start again? Probably not. If the magnet is strong enough to lift the ball all the way from the bottom to the top, it's also strong enough to keep the ball from ever reaching the bottommost point so it can start the trip over again.

Whether or not the machine works as I describe, I'm sure it won't work for another reason, and that's conservation of energy. In the real world, there has to be at least a tiny amount of friction. That means this system is continually losing energy to thermal energy. So there is less and less energy available for the kinetic energy of the ball. But the ball can't keep losing kinetic energy each trip, or the machine will stop running. The only way to keep this machine working indefinitely is to continuously input energy, which you might do as follows.

When the metal ball reaches the top of the ramp and falls through the hole, you remove the magnet from the system. This allows the ball to return *all* the way to the bottom of the apparatus. Then you replace the magnet, which draws the ball up to the top again. By continually removing and replacing the magnet, you keep the ball moving. With the act of removing and replacing the magnet, though, you are adding energy to the system (you do work on the magnet when you move it). If you have to put energy into the machine, it's not a perpetual motion machine.

One point of clarification. I say perpetual motion machines won't work because they violate the principle of conservation of energy. But what if someone builds a perpetual motion machine that actually works? That would mean conservation of energy is incorrect! This principle has been around for a very long time, and so far no one has found a case where it doesn't work, but that doesn't mean it will never happen. As with all science concepts, you have to realize that people *made up* conservation of energy. Science concepts are really great things but they're not chiseled on stone tablets. If your neighbor produces a perpetual motion machine and isn't scamming you (put your money on the scam), then we'll all have to change our thinking about conservation of energy.

Chapter Summary

- Simple machines are devices that allow someone to trade an increase in distance moved for an increase in applied force, or vice versa.
- By applying conservation of energy to the operation of a machine and ignoring certain kinds of "lost" energy, you can determine the relationship between the forces and distances you input to a machine and the forces and distances that are output from the machine. That relationship is $F_1d_1 = F_2d_2$.
- Because energy losses in machines are a reality, we speak of the *efficiency* of a machine, which can be calculated using the formula:

$$\text{Efficiency} = \frac{\text{work out}}{\text{work in}} \times 100\%$$

- If you believe in conservation of energy, then perpetual motion machines are physically impossible to build. If you believe in fairies, clap your hands.

Applications

1. Here are a few common levers: fingernail clippers, bottle openers, and lock cutters (they use these at the local YMCA to cut locks off lockers when people forget their combinations). Take a close look at a set of fingernail clippers (Figure 3.29). As you push down on the upper arm, the blades of the clippers come together. The mechanism for this is a bit more complicated than a ruler and a pencil but it's the same basic idea. You are trading the extra distance the upper arm moves for the extra force you get on the cutting end of the clippers.

Figure 3.29

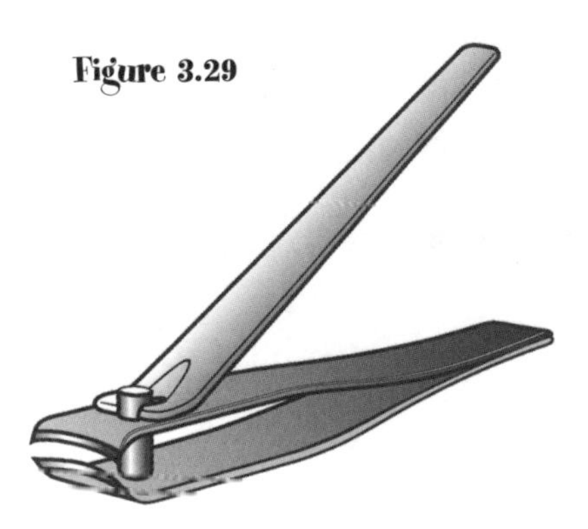

Now that you know how fingernail clippers work, explain why toenail clippers are so much larger. Hint: It has to do with increasing the force more than is possible with fingernail clippers.

Bottle openers are another example of a simple lever. You move the handle of the bottle opener much more than the bottle end of the opener moves. Trading distance for force.

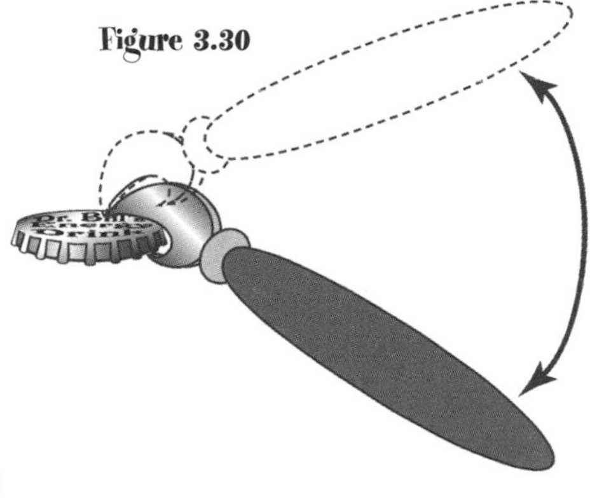
Figure 3.30

As for the lock cutters, just look at Figure 3.31 to see how they work. Consider it a homework assignment!

Figure 3.31

2. Ever wonder how car jacks allow you to lift an entire car off the ground? Wonder no more. I'll explain hydraulic car jacks here. Other kinds of car jacks use this same basic idea but with different mechanisms. Figure 3.32 shows a standard "floor jack," also known as a hydraulic jack.

Figure 3.32

To operate this kind of jack, you put it under the car at a place where you won't damage the car and you start pumping the handle. You pump the handle a lot, and the car rises just a little. This is our well known distance-for-force trade-off. Figure 3.33 shows a hydraulic system that is a lot like a hydraulic car jack.

Figure 3.33
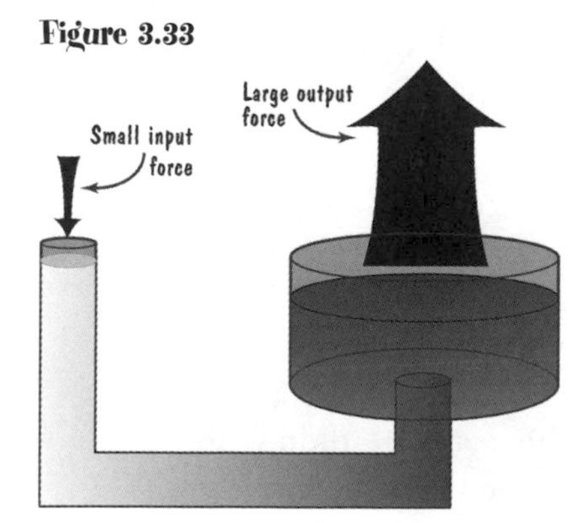

When you push down on the left side of this system, it pushes the liquid in the system so the liquid pushes up on the right side of the system. Liquids are special in that they don't compress much, if at all, when you push on them. So what happens is that pushing a large distance on the left side makes the right side go up just a little. This is because the two cylinders have such different diameters. The total force exerted on the right hand side is correspondingly much larger than the force exerted on the left-hand side. In fact, the force is large enough to lift a car. Log splitters, backhoes, and many other pieces of heavy equipment use hydraulics in this way to increase the force you can apply to something.

3. How do those cranes lift very heavy objects such as metal beams and big slabs of concrete? Pretty simple. Cranes use a basic pulley system, much like a block and tackle, to trade distance for force.

4. One basic kind of machine I didn't address in this chapter is the machine known as the **gear**. Below is a drawing of two gears.

Figure 3.34

When you turn the gear on the left, it causes the gear on the right to turn in the opposite direction. Because of the different sizes of the gears, the one on the right is also able to exert a greater force on something else than you are exerting on the handle. It's a force-distance trade-off again. The gear on the right doesn't move as far as you move the handle on the left, but you get an increase in force.

And just so you don't go away with a big ol' misconception, what I just said about distance and force is true only because the handle on the left gear moves in the same-sized circle as the edge of the gear on the right. If these were different sizes, I would have to talk about that thing called torque to be accurate. And that is the reason I didn't discuss gears in the main part of the chapter! For a great example of gears in action, check out a 10-speed bicycle and try to figure out why the various gears (they're called sprockets on a bike) are the sizes they are.

SCI LINKS. THE WORLD'S A CLICK AWAY

Topic: simple machines

Go to: *www.sciLINKS.org*

Code: SFE10

Temp-a-chur and Thermal Energy

Yeah, yeah, I know it's supposed to be temperature, but the above spelling is how my kids used to pronounce it and it was just so darned cute! Anyway, I've been using the term *thermal energy* for three chapters now, so maybe it's time to get a better picture of what that is. Before we get to that, though, let's explore what we mean by temperature.

Things to do before you read the science stuff

Set out three cups. Put cold tap water in one, room temperature water in another, and hot tap water (not so hot you can't put your hand in it) in the third. Place your left hand, or just the fingers if it's a small cup, in the cup of cold water. At the same time, place your right hand in the cup of hot water. Keep your hands there for about a minute. Then alternately stick your left and right hands into the cups of room temperature water. This water will feel warm to your left hand and cool to your right hand. Neat, huh?

The science stuff

Obviously, humans don't make very good thermometers. How hot something feels to you depends on how hot or cold your skin happens to be. Might be nice to have a more reliable way to determine temperature. Before doing that, let's agree that temperature is nothing more than a measure of hotness or coldness. That's a nice, simple definition but it's also pretty vague. To make it more precise, we need a standard definition that everyone can agree on. To develop that standard, it's useful to look for physical things that seem to occur over and over again at the same degree of hotness. In other words, they occur at the same *temperature* each time.

It so happens that two such occurrences are quite common: when water freezes or melts and when water turns to steam (otherwise known as the boiling point of water). Under tightly controlled conditions, these two things consistently occur at two specific temperatures. What scientists have done, then, is arbitrarily assign numbers to these two temperatures in order to develop temperature scales—entire ranges of numbers that assign a number to each temperature we experience.

Unfortunately, we have three different temperature scales to deal with. In the Fahrenheit scale, water freezes at 32 degrees and boils at 212 degrees. In the Celsius scale, water freezes at 0 degrees and boils at 100 degrees. In the Kelvin scale, water freezes at 273.15 degrees (!) and boils at 373.15 degrees. And in case you're wondering, all three scales are named after the physicists who made them up. So if you want your name to be immortal, come up with a new temperature scale and convince the scientific community to use it.

Figure 4.1

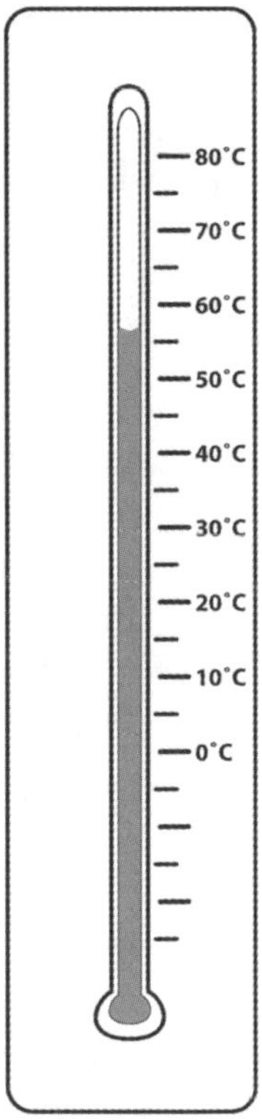

80°C
70°C
60°C
50°C
40°C
30°C
20°C
10°C
0°C

Speaking of which, scientists use only the Celsius and Kelvin scales in their work. The Kelvin scale is the official SI unit for temperature, and the intervals on the Celsius scale are the same size as those on the Kelvin scale, so it's easy to convert from one of those scales to the other. Converting from those scales to the Fahrenheit scale, though, is a bit more complicated, as you might recall from your seventh grade science class. Or maybe you'd rather forget that experience.

More things to do before you read more science stuff

Find yourself a mercury or alcohol thermometer—one that measures temperature by the change in height of a column of liquid.[1] (See Figure 4.1.)

Take your thermometer and traipse around finding the temperature of various things. You might start with those three cups of water. Watch as the column of liquid rises and falls,

[1] Mercury thermometers are difficult to find these days, simply because mercury vapor is so poisonous. Before people knew mercury was dangerous, my mom (a nurse) used to let my siblings and me play with the mercury that spilled from a broken thermometer. My wife says that explains a whole lot about how my brain works.

depending on how hot something is. Why in the world does the column of liquid do that?

While pondering that question, head to the kitchen sink. Turn on the cold water and let it run until the water is as cold as it's going to get. Then gradually decrease the water flow until it's just a tiny stream. Wait a minute or two and see if you notice any change in the size of the stream. Now switch over so you have a tiny stream of hot water. Make sure you don't turn the hot water on full force as you do this. Just wait for the tiny stream to gradually get hot as the hot water replaces the cold water. As this happens, you should notice the stream getting narrower, maybe even dwindling to a trickle or shutting off altogether. If that doesn't happen, run the cold water full force for a minute and try again with the tiny stream of hot water.

One last, simple thing to do in this section. Put glasses of cold and hot water side by side. Put a drop of food coloring in each and watch what happens. Pay special attention to the time it takes for the food coloring to disperse in each glass.

More science stuff

At an early age, we all learn that everything in the world is made of tiny things called molecules. Before using molecules to explain what you observed, I'm going to try and put the concept of molecules into perspective. The concept that molecules exist, and descriptions of how those molecules behave, is an example of what's known as a **scientific model**. Scientists use models when they can't directly observe something. We make up a picture of how it would look if we could shrink down small enough and look at the molecules directly, something we can't do yet, except in science fiction movies.

Many years ago, scientists came up with a molecular model and showed how it could explain a whole bunch of observations. As more and more evidence piled up that was consistent with the existence of molecules, people became more and more convinced that the model was a good one. Today, molecular theory is such a reliable model that scientists accept as fact the existence of molecules and the rules for how they interact. Keep in mind, though, that it's not the same kind of directly observable fact as "There are four people in my house." It's more like the following statement of fact: "I have not directly seen four people in my house, but everything else I see and hear is consistent with there being four people in my house."

At any rate, the model states that the molecules of any substance are in constant motion. In a solid, they vibrate and possibly rotate a bit and in a liquid, they move and slide across one another in addition to vibrating and rotating.

The food coloring in water is evidence that the molecules of water are in motion. If they weren't in motion, you'd expect the food coloring to just drop to the bottom and not spread out. But the food coloring spreads out pretty fast. That's because the water molecules in motion bump into the food coloring molecules, which also are in motion, spreading them throughout the liquid. The hotter the water, the more rapidly the food coloring disperses. We take that to mean that hot water molecules move faster than cold water molecules.

This leads us to the basis of the concept of temperature and what it means for one thing to be hotter than another. The hotter something is, the faster its molecules are moving. And that means that a hot collection of molecules has more *kinetic energy* than an equal number of cold molecules. We can take that one step further and say that the more kinetic energy possessed by the molecules in an object, the more *thermal energy* the object has.

Topic: thermal energy

Go to: *www.sciLINKS.org*

Code: SFE11

All right, what's going on with that column of liquid in a thermometer? Suppose you put the thermometer in a cup of hot water. All those water molecules are bouncing around at relatively high speeds because the water is hot. The molecules in the thermometer liquid start at room temperature so they're moving quite a bit slower. When the two come in contact,[2] the fast-moving molecules collide with the slow-moving molecules. In these collisions, the fast-moving molecules transfer some of their kinetic energy to the slow-moving molecules, making the slow ones move faster than they were. The end result is that the liquid in the thermometer is now hotter than it was.

So the molecules in the thermometer liquid are moving faster. So what? Here's what. As they move faster and faster, they push harder on each other and their surroundings and, if not confined, they take up more space as they pull away from one another. What that means is that the liquid expands up into the open space in the tube. The liquid also contracts when it gets cooler because there are forces between the molecules that tend to pull them together.[3]

Solids behave a whole lot like liquids when you heat them. When a solid gets hotter, its molecules vibrate faster and tend to pull away from one another. This makes the solid expand. Just about every solid expands when heated and con-

[2] There's actually glass or plastic in between the two liquids. All that means is that there's a "middleman" in the transfer of energy that I'm about to discuss.

[3] The forces between molecules are electrical forces. Unfortunately, a discussion of those forces would take us too far off track at this point.

tracts when cooled.[4] That information gives us the explanation for what happened with the kitchen faucet. The valve in the handle(s) is made of metal. When you switch from cold to hot water, the metal in the valve expands and, with most designs, actually closes the valve a bit. If you have a very small stream of water, this metal expansion can noticeably reduce the flow of water and even shut it off. You don't notice any change with the valve open more than a little because of the larger volume of water flowing out.

Let me recap before going on to the relationship between temperature and thermal energy. Both liquids and solids expand and contract with changes in temperature. The molecules that make up these liquids and solids are in constant motion. The hotter a substance, the faster its molecules move, and this explains the expansion and contraction of solids and liquids.

Now for the relationship between temperature and thermal energy. In previous chapters, I focused primarily on thermal energy being the kinetic energy of molecules that results from friction. More completely, thermal energy is a measure of the total kinetic energy possessed by an object plus any potential energy the molecules have because of their relative positions or individual shapes. Temperature, depending as it does on how fast molecules are moving, also is related to the kinetic energy of the molecules. The two are different, though, as the following example illustrates.

Suppose you have a metal rod that's at a temperature of 30 degrees Celsius. No matter what point on the rod you choose for measuring its temperature, you will get 30 degrees for your answer. Now suppose you cut the rod in two. What are the temperatures of the two halves? Still 30 degrees, right?

Figure 4.2

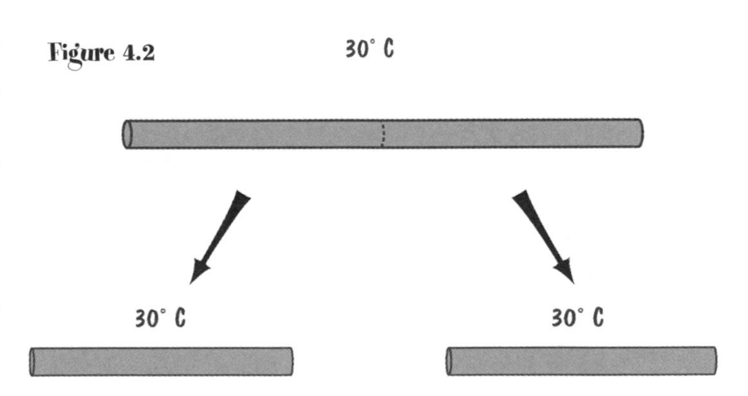

[4] One exception is water, which actually expands as it cools in the very limited temperature range between 0° Celsius and 4° Celsius. The reason it expands when cooling in this temperature range has to do with the particular way water molecules latch onto one another as they cool. This peculiar expansion of water means that just as water is about to become ice (at 0 degrees Celsius), it becomes less dense than the water around it. That makes it float on top of warmer water and is responsible for the fact that ice forms on the *surface* of ponds and lakes rather than sinking down and forming at the bottom—a big deal if you happen to be a fish that wants to survive the winter!

Now consider the total thermal energy contained in the uncut rod. Just add up the kinetic and potential energy of all the molecules in the rod, and you have your answer.[5] When you cut the rod in half, each piece, having half the number of molecules as the original, now has half the original amount of thermal energy.

Figure 4.3

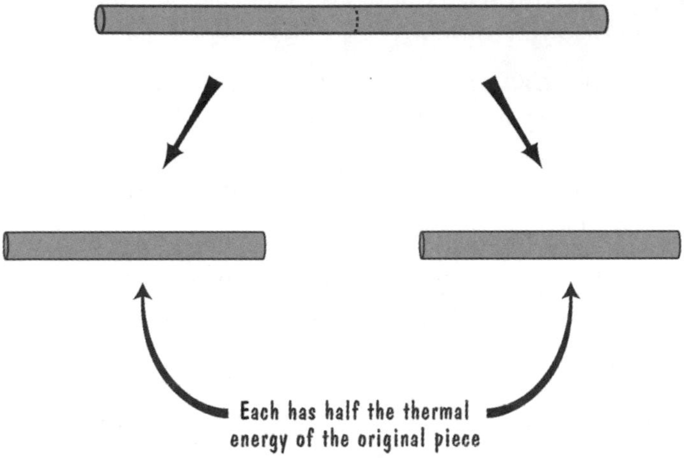

Each has half the thermal energy of the original piece

Thermal energy is related to kinetic energy, but temperature and thermal energy clearly aren't the same thing. Temperature has more to do with the *average kinetic* energy of the molecules, while thermal energy is a measure of the *total* energy of the molecules. With those definitions, it's possible for a cold object to actually have more thermal energy than a hot object. A swimming pool full of cold water has much more thermal energy than a thimbleful of hot water, simply because the swimming pool has so many more water molecules.

Figure 4.4

[5] Thermal energy is also known as *internal energy,* and it turns out to be an extremely complicated task to add up all those kinetic and potential energies. Not only that, but something called *quantum mechanics* comes into play at such a small scale, and the simple molecular model I'm presenting falls apart. That simple model is just fine, though, for explaining what we're doing in this book.

Even more things to do before you read even more science stuff

Figure 4.5

Blow up a balloon and tie it off. Measure the diameter of the blown-up balloon.[6] Try to measure at least to the nearest centimeter. Now get hold of an empty 12-ounce pop bottle, beer bottle, or water bottle (plastic and glass both work). Find a coin that will cover the top of the bottle completely (no cap on the bottle). A dime works well for a beer or pop bottle and a quarter works for water bottles that have larger openings.

Put both the balloon and the empty bottle in the freezer. Keep the coin in your pocket. Wait at least a half hour (have another bottle of, er, —water), remove the bottle from the freezer, and set it on a flat surface. Wet one surface of the coin and place it, wet side down, on top of the cold bottle. Make sure the coin completely covers the opening. Watch closely for a while. Anything strange happen? It should!

Once the coin stops doing its thing, get the balloon out of the freezer and measure it again. Any change from the earlier measurement? Wait about a half hour and measure the balloon again. It should be back to its original size.

Even more science stuff

In the previous section I dealt with what happens to solids and liquids when they change temperature. In this section we're dealing with gases—specifically, just plain ol' air. To get us started in the explanation, I'm going to explain one simple scientific model for what gases are. Basically, just picture a whole bunch of very tiny molecules flying around in space and bumping into one another. The only way these molecules affect one another is when they have a collision. When they collide, they might transfer energy from one to another, but they don't lose energy to sound or friction or anything like that. These are not like collisions between cars—more like bumper pool for itty-bitty things.

As with solids and liquids, the faster gas molecules move, the higher the temperature of the gas. Also like solids and liquids, the temperature of a gas is directly related to the average kinetic energy of the molecules of the gas. If you double the average kinetic energy of a collection of gas molecules, you double the temperature of the gas. There's a whole lot more to this model of what a gas is, but we have enough now for our purposes. And by the way, this model is

[6] I'm assuming you have a round balloon. If you have a hot-dog shaped balloon, measure its width and length.

Figure 4.6

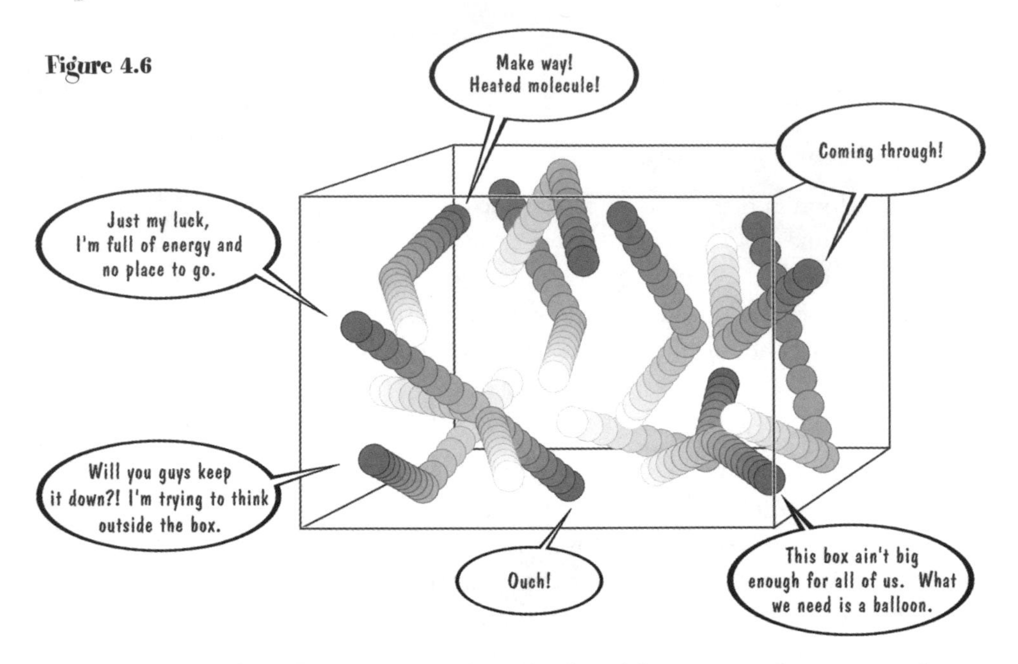

known as the **kinetic theory of gases.**[7] Feel free to use that term to impress friends and enemies alike.

Time to explain what you did in the previous *Things to do* section. I'm going to start by giving you the *wrong* explanation for one of the activities. My reason for doing this is that the wrong explanation shows up in way too many textbooks. I want you to be on the lookout for it. And don't worry—I'll give you the correct explanation afterwards. In the meantime, here's the wrong explanation.

> ***Incorrect explanation:*** *When gas molecules are moving at relatively slow speeds, they don't need much space. When you heat those molecules, they move much faster and require more space. Therefore, gases expand when you heat them. When you cool a gas down, the molecules don't need as much space, so the gas as a whole contracts.*
>
> *We saw how that works in the balloon activity. Put a blown-up balloon in the freezer, and the gas molecules inside slow down. They don't need as much space when they're moving slowly, so the balloon collapses a bit. Bring the balloon back out to room temperature and the air molecules inside heat up, move faster, and require more space. That's why the balloon gets bigger.*

Okay, so much for the incorrect explanation. What's incorrect about it is that fast-moving gas molecules *do not need any more space* than slow-moving gas

[7] I want to stress that this is a *simplified* model. We're not taking into account the fact that molecules do have size, that they rotate and vibrate, and that their collisions aren't 100 percent like billiard ball collisions.

molecules. Also, gas molecules *do not necessarily expand when you heat them*. To demonstrate this, I'm going to describe an activity you can do with roughly ten or more people. If you have a classroom full of kids to use, great. If not, use this activity to spice up your next social soirée. Beats charades. Create some floor space by moving furniture or head outside. For ten people, you'll need an area about 4 meters by 4 meters. Mark off the boundaries of this area. It can be any shape but I'll use a square in this explanation. Have your people stand inside the boundaries.

Figure 4.7

Figure 4.8

The people are going to pretend to be gas molecules, and the boundaries represent walls that keep the gas molecules confined. At your signal, all of the people-molecules are to begin moving according to the following rules.

- Move in a straight line until you collide with a wall or another gas molecule. In between collisions, always walk in a straight line at a constant speed. Don't alter your path in order to avoid or to collide with another molecule.
- When you collide (gently!) with a wall or another molecule, bounce off as if you were a billiard ball.

- Keep your hands to yourself—no groping! This rule is especially important for adolescents!

Have the people start by pretending to be a cool gas; they should move about slowly. Gradually have them get hotter (speed up). Stop them before the collisions get too violent. Having observed these people bumping around, you should be able to answer a few questions, no? Whether you're ready or not, here are the questions. No need to be nervous about this test—answers are provided.

1. Did the gas expand when it got hotter? Answer: No. The walls are rigid, so the molecules were confined to the same space whether they were hot or cold.
2. Do cool gas molecules use all of the space provided? Answer: Sure. It might take them longer to cover the entire space than it would hot gas molecules, but they would eventually cover the entire area. Suppose the people-molecules had paint on their shoes. If you wait long enough, the entire room would be painted, whether the people are moving quickly or slowly.
3. Is the space between the molecules any different when the molecules are moving quickly as opposed to when they're moving slowly? Put another way, could you tell from an aerial snapshot (fast shutter speed, so no blurring) whether the molecules were moving quickly or moving slowly? Answer: No. Any aerial snapshot would just show 10 bodies distributed around the available space. Hot gas molecules don't have any more space between them, on average, than do cold gas molecules.

In summary, hot gases don't *require* any more space than cold gases, and heating a gas doesn't necessarily cause it to expand. When you cool a bunch of gas molecules, they move more slowly but continue to take up the entire space afforded them. They don't huddle in a smaller space.

Fine. Now what about that balloon? It *did* expand and contract when heated and cooled, and so did the air inside. This is because the "walls" of a balloon stretch. When you heat the air inside a balloon, the molecules move faster, hitting one another and the inside of the balloon much harder and more often. What results is an overall greater push from the inside, and the balloon stretches outward. When the air inside cools, the molecules inside don't hit the balloon as hard or as often. The balloon then naturally contracts to a smaller shape. That leads us to the *correct* statements about how gas molecules behave.

Correct explanation: *If the container holding a gas* allows *it, heating a gas will cause it to expand. If the container holding a gas naturally tends to contract its shape, then cooling a gas will cause it to contract.*

Now that you're a gas expert, let's look at the coin and bottle. Keeping the bottle in the freezer for a while ensures the air inside the bottle is cold, resulting in relatively slow-moving molecules. When you bring the bottle out into the room, the air molecules inside the bottle *itself* start to move faster. Why? Because the fast-moving air molecules in the room bombard the bottle, causing the molecules in the bottle itself *(glass or plastic molecules)* to vibrate faster. This faster vibration causes the air molecules inside the bottle to speed up. It's a simple transfer of kinetic energy.

By putting the coin on top of the bottle, you confine this gas (the air inside the bottle) and keep it from going anywhere. The water you put on the coin helps create an airtight seal. As the air molecules inside the bottle heat up, they move faster and faster, meaning they hit the coin harder and more often. Eventually, they hit the coin hard enough to make it jump up. When the coin jumps, air molecules inside the bottle escape.[8] Now you have fewer air molecules inside than before, so the coin doesn't get hit as often.

But the air molecules inside the bottle aren't finished heating up. As they get hotter from contact with the outside air, they move faster than before and finally push the coin up again. More molecules from inside the bottle escape. This process continues until the molecules inside the bottle reach room temperature, or until there just aren't enough air molecules inside the bottle to push up the coin, no matter how fast they might be moving.

Chapter Summary

- With a few exceptions, solids and liquids tend to expand when you heat them and contract when you cool them.
- Gases expand when heated if they are allowed to do so and contract when cooled if their surroundings are able to contract around them.
- The temperature of an object or substance is directly related to the average kinetic energy of the molecules in the object or substance.
- Thermal, or internal, energy is a measure of the total kinetic and potential energy within an object. Thermal energy and temperature are not the same thing.

[8] Simply stating that the molecules inside the bottle escape is a bit simplistic. Actually, some outside air molecules enter the bottle while those inside are leaving. Because the cold air inside the bottle is actually denser than the outside air, however, the overall effect is for more air molecules to leave the bottle than enter it.

Figure 4.9

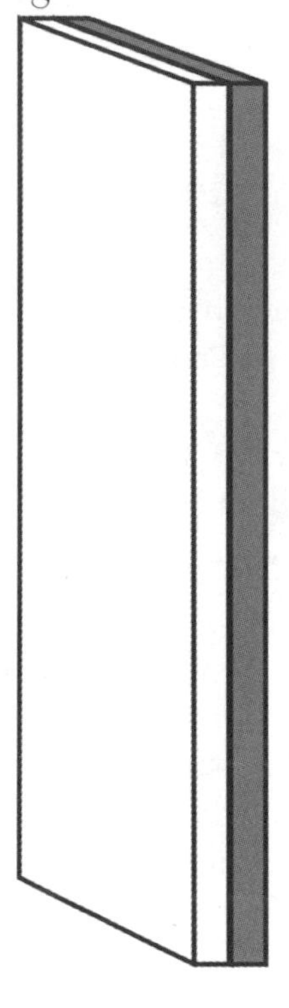

Applications

1. People who build bridges, railroad tracks, and tall buildings know all about expansion and contraction of metals. If you build a bridge out of very long, solid rods of steel, you're asking for trouble. When the outside temperature changes dramatically, these long sections of steel will buckle and bend due to the expansion and contraction of the metal. The solution? Make your structure with many shorter pieces of metal with gaps in between to allow for expansion and contraction due to temperature changes. Of course, if you live near the equator and the temperature doesn't vary much, go ahead and use those very long sections of metal.

2. Different kinds of metals expand and contract different amounts in response to a given change in temperature. Knowing that, clever people have invented what are known as bimetallic strips. These are just two strips of metal that are fastened together, as in Figure 4.9.

   When you heat this bimetallic strip, one side expands more than the other side. To see what this is like, put your hands together, as in Figure 4.10.

Figure 4.10

   Keep your fingertips together but push up with one of your hands. This is like one side of the metal expanding faster than the other side. What happens is that your hands bend in one direction (see Figure 4.11).

Figure 4.11

   Well, that's neat, but what does it have to do with anything? To answer that, go take the cover off the thermostat in your house or apartment. Somewhere in there you'll see a coiled metal band. That coil is actually a bimetallic strip. When the temperature changes, the difference in expansion and contraction between the two metals in that strip causes the coil to wind or unwind. This triggers a switch (usually a tube containing mercury) to turn your furnace or air conditioning on and off.

Figure 4.12

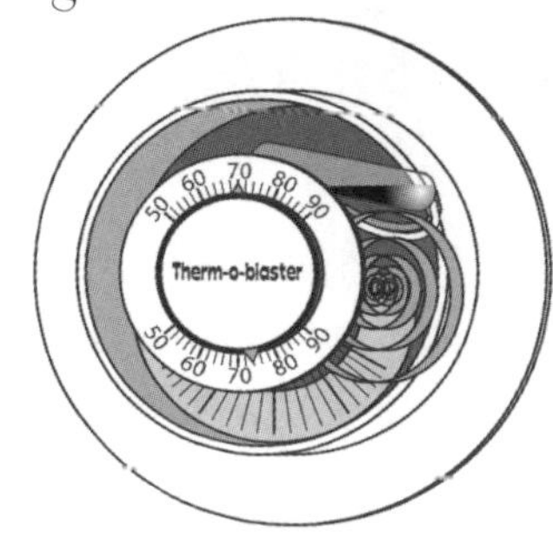

   If you have a dial-type thermometer (indicator of temperature goes in a circular path), check out the back of it and you'll find the same coiled bimetallic strip.

3. We don't have quite enough background to completely explain how hot air balloons work, but we can address one component. One reason hot air balloons rise is that the hot air molecules inside the balloon are farther apart than the cooler air molecules outside the balloon.

But doesn't that contradict what I told you in the last section? Nope. Hot air balloons have at least one opening at the bottom. As the air inside the balloon gets hotter, the molecules move faster and push harder on one another. This forces a bunch of them out the opening, leaving fewer air molecules inside than before. With fewer of them, the molecules are farther apart than before. Of course, these hot air molecules inside push harder on the balloon walls than would cooler air molecules. Even though there are fewer of them, their greater push keeps the balloon inflated.

What do you think would happen if the balloon operator kept firing up the burner and continually heating up the air inside? Would you get to a point where you would have pushed almost all the air molecules out of the balloon? And if you could do that, wouldn't the balloon then collapse, with little or no air inside? Well, first of all, yes; if you could somehow get rid of most of the air inside the balloon, the balloon would definitely collapse. However, you could never get to that point. There are limits to how hot you can make the air inside the balloon. At some point you'd reach an equilibrium where the less dense, hotter air inside the balloon would push out just as hard as the more dense, cooler air molecules outside the balloon push in. After that point, you couldn't force any more air out of the balloon.

Close the Door–You're Letting the Cold In!

Actually no, you aren't letting the cold in when you leave the door open in winter. That's a common misconception that we will gently put to rest in this chapter. We will, however, be discussing the transfer of energy that makes things hotter and colder.

Things to do before you read the science stuff

Hold your hand in some warm water, first making sure the water isn't so hot you burn yourself (Doh!). Your hand feels warmer, right? How does the energy that makes your hands warmer get to your hands?

"No, Mom. I'm not letting the cold in. I'm letting the heat out!"

Go outside on a sunny day. First stand in the shade and then in the sun. Where do you feel warmer? If you're in Phoenix in the summertime, you feel hot no matter what and if you're in Minneapolis in the winter, you feel cold no matter what. Just do your best to find a difference. And it's not like you don't know what the answer *should* be! Now answer a question. How did that energy that made you warmer in the sunlight get to you?

Go back inside and find a table lamp. Take off the lampshade and turn on the light. Hold your hands near the bulb (don't touch—it burns!). They feel warmer, yes? How does the energy that makes your hands warm get to your hands?

Time to show that a watched pot does indeed boil. Get a large diameter cooking pot (best if it's larger in diameter than the heating element you plan to put it on) and fill it at least half full of cold water. Have food coloring ready, along with something that will float on top of water. I used crumpled-up potato chips for the latter (check under the sofa cushions). Once you have everything, put the pot on the stove and turn on the heat. After a few minutes, put a squirt of food coloring in the water near the edge of the pot. What kind of path does the food coloring take? Wait a few more minutes and add another squirt of food coloring. Add more food coloring in different parts of the water. Try to detect any pattern in the motion of the water. Finally, put your floating stuff on the surface of the water and watch for patterns. Why does the water move horizontally? What's causing that?

SCiLINKS. THE WORLD'S A CLICK AWAY
Topic: heat and temperature
Go to: *www.sciLINKS.org*
Code: SFE12

The science stuff

In each of the things you just did, there was a transfer of energy. You know that because in each case something gained thermal energy (got warmer). The Sun made your body feel warmer, as did the light bulb. The warm water also made your hand feel warmer, and the burner on the stove caused the water to get warmer. For all these situations, we say that an amount of **heat** was transferred from one thing to another. Heat is the amount of energy transferred from one place to another because of temperature differences.[1] As such, it is impossible for an object to *possess* heat or heat energy. The energy is only called heat when it's going from one place to another. Because heat is energy, it's measured in

[1] There's a natural tendency for heat to travel from hot objects to cold objects via the mechanisms I'm about to describe. This tendency is part of a general tendency for everything in the universe to reach a state of equilibrium.

joules. Usually, however, you'll see heat measured in *calories,* which are not exactly equal to joules, but are still a unit of energy.[2]

It should be no surprise to you that in all these transfers of heat, conservation of energy still applies. In fact, when you apply conservation of energy to heat transfer, you have a special "law" known as the first law of thermodynamics. Here's what it looks like as an equation:

(Heat added to a system) = (change in thermal energy of the system) + (work done by the system)

Before trying to make sense of this, let me remind you what a system is. It's any object or collection of objects around which you have an imaginary boundary. When you apply conservation of energy, you always have to specify what system you're talking about. And if you'll think back to Elroy and his blocks (Chapter 2), you'll remember that if you don't have a *closed* system, you have to keep track of any energy that enters or leaves the system. Now, because the first law of thermodynamics deals with heat entering the system and work done *by* the system, we're obviously not talking about a closed system.

Okay, in this law, you're adding energy to a system and you're doing it through heat. What could possibly happen to that added energy? Well, it could show up as an increase in the thermal energy of the system, which means an increase in the sum of all the internal kinetic and potential energy of the system. This increase in thermal energy can do work on other things and thereby cause energy to leave the system. For example, gas molecules bouncing against a movable wall can cause it to move. And basically, that's it. If you keep track of the energy added to the system (that's the "heat added" term), the energy that stays in the system (that's the "change in thermal energy" term), and the energy that leaves the system (that's the "work done by" term), you'll find that the first law of thermodynamics holds true. In symbol form, that equation reads:

$$Q = \Delta U + W$$

Q is the symbol physicists use for heat, U represents thermal (or internal) energy, and W represents work. Δ is the Greek letter delta and it means "change in." We're not going to go any further with the first law of thermodynamics, but I did want to show it to you in case you come across it sometime. If and when that happens, I hope you remember that it's just a statement of conservation of energy when heat is involved.

Topic: thermodynamics

Go to: *www.sciLINKS.org*

Code: SFE13

[2] To confuse things even further, the kind of Calorie associated with food intake isn't the same as the calorie used in scientific calculations!

So, heat goes from place to place and when it does, it obeys conservation of energy. Story's not over yet, though. It turns out there are several ways heat can go from one place to another, and it's useful to describe and name those different ways. You experienced all three if you did everything I asked you to in the previous section. First, you put your hand in warm water. That brought the molecules in your hand in contact with fast-moving water molecules. As those fast-moving water molecules bombarded your hand, they transferred energy to the molecules in your hand, giving your hand an increase in thermal energy. In other words, heat went from the water to your hand. When heat transfers by direct contact between molecules, that's called **conduction**.

Figure 5.1

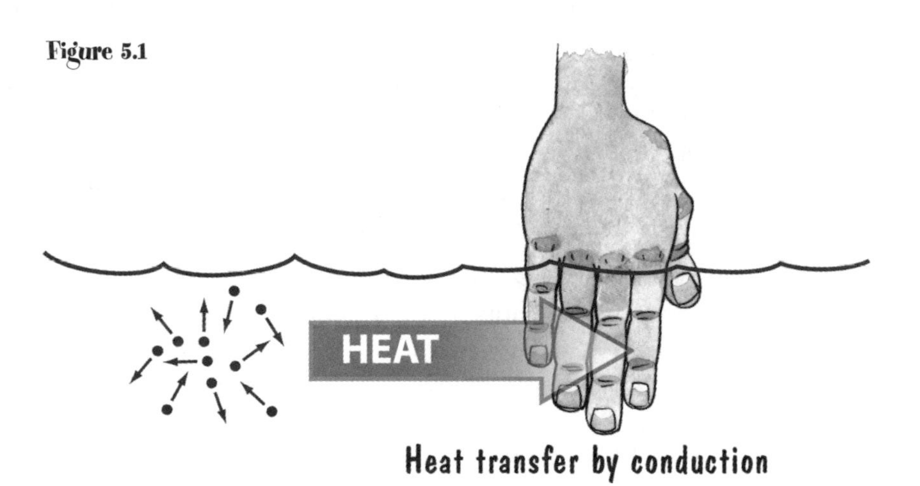

Heat transfer by conduction

Next you went outside and let the Sun warm you up. It's a pretty good bet that the Sun didn't touch you, so that couldn't have been transfer of heat by conduction. The space between the Earth and the Sun is basically empty, so there aren't even any "middlemen" that might help conduction occur. The Sun transferred heat to you through **radiation**. In addition to putting out visible light, the Sun puts out invisible light. The invisible light is basically the same stuff as the visible

Figure 5.2

"I knew Florida was a bad choice for the company picnic."

Heat transfer by radiation

light but it has slightly different properties.[3] One thing both visible and invisible light have in common is that they can go from one place to another without any molecules or anything else in between. In other words, they can travel through empty space.

SCiLINKS. THE WORLD'S A CLICK AWAY
Topic: radiation
Go to: *www.sciLINKS.org*
Code: SFE14
Topic: properties of light
Go to: *www.sciLINKS.org*
Code: SFE15

One kind of invisible light, called **infrared** light, heats things up. And that's the kind of radiation that made you feel warm when standing in sunlight.

Light bulbs also put out infrared radiation, so that's part of what heats up your hand when you put it near a lit bulb. This situation also involves conduction, though. The air molecules next to the bulb absorb heat by being in contact with the light bulb. They transfer this energy to other air molecules, which in turn transfer heat to your hand by conduction (Figure 5.3).

Figure 5.3

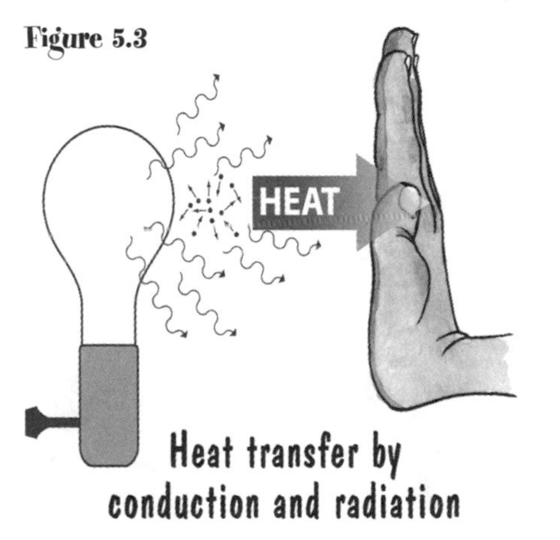

Heat transfer by conduction and radiation

The third type of heat transfer I'm going to talk about isn't really heat transfer, but since it does involve hot and cold molecules moving from place to place, it seems to fit in here.

When you put food coloring into water that was getting hotter and hotter, you should have noticed that the food coloring didn't just spread out as it would in a glass of cold water. If you put the food coloring near the edge, chances are it dropped straight down and then went towards the center of the pot once it got to the bottom. If you put it in the center, it most likely moved horizontally to the edge and then dropped down. Whatever floating things you put on the surface also should have showed you that there is a definite pattern in which water molecules move horizontally across the surface of the water. This is because the water in the pot rises in some spots, falls in others, and moves horizontally in others. From a side view, it might look like Figure 5.4.

Figure 5.4

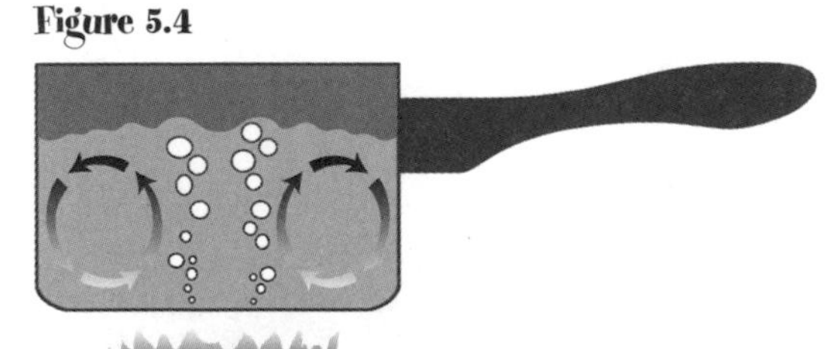

[3] For more on this, see the *Stop Faking It!* book on *Light*.

That drawing assumes that you were heating the water just in the center of the pot, and that the pot was much larger than the heating element. If you had a different situation, you probably observed much more complex patterns than what is drawn here. At any rate, here's what's going on. When you heat water at the bottom of the pot, those molecules move faster and push against other water molecules, creating extra space for themselves.[4] Cooler molecules at the top of the water are closer together. Molecules of a liquid that are more closely packed than other molecules of a liquid, will push down and displace the less closely packed molecules, which are pushed up (Figure 5.5).

Figure 5.5

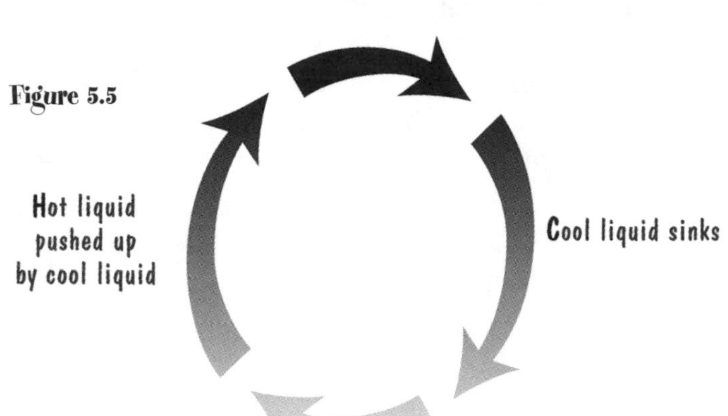

This process is known as **convection**. As cooler water sinks down, usually around the edge of the pot, it pushes hotter water in the center up to the top. That cooler water then heats up and water cooler than it on the edge pushes *it* up to the top. This circulation keeps going, creating what's known as a **convection cell**.

SciLINKS
THE WORLD'S A CLICK AWAY
Topic: convection
Go to: *www.sciLINKS.org*
Code: SFE16

In this way, hot water molecules and cold water molecules circulate and end up distributing hot water throughout the pot. This also works for air molecules. And before I go on, I have to say something quick about the phrase "hot air rises." Yes, it does, but not without the help of cold air pushing it upwards. Hot air doesn't rise all by itself, but I guess that discussion will have to wait for another book.

Figure 5.6

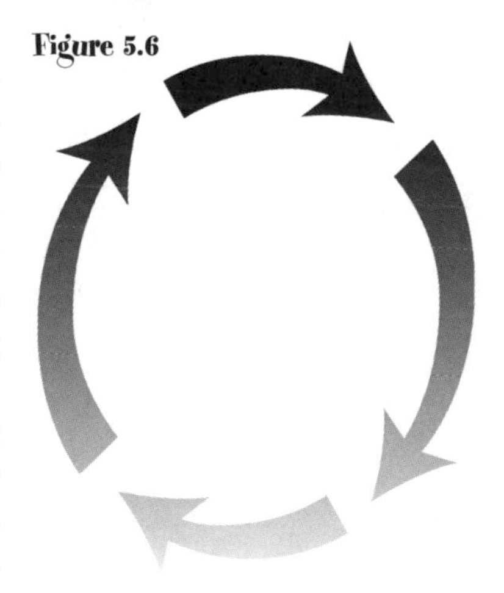

A convection cell

In summary, you can transfer heat by conduction or radiation, and hot water or air can move from

[4] This is a case where the water is allowed to expand because it's pushing on other water molecules, which are able to move.

place to place through convection. When talking about the transfer of energy, with no large-scale movement of hot or cold air or water, it is *heat* that travels from one place to another, and not cold. So when your house gets colder in winter because you have only single pane windows, you are letting heat out. When your house gets warmer in summer for the same reason (single pane windows), you're letting heat in. You are never letting cold in or out. You can let *cold air* into or out of your house, but you can't let *cold* in or out.

More things to do before you read more science stuff

Still have that pot of really hot water around? If not, heat up another. Also get a screwdriver, a wooden spoon, and a sheet of aluminum foil. Form the aluminum foil into a long, thin shape. Remove the pot from the heat and place these three things in the hot water, as shown in Figure 5.7.

Figure 5.7

Touch each of the objects just above the water line and see how hot each feels. Then do the following with each object: Remove the object, dry it with a towel, and then touch the part that was in the water. Note how hot each feels.

Fill a glass with ice and then add cold tap water to about the three-quarter point of the glass. Let it sit for a while. Use a thermometer to measure the temperature of the ice water, making sure the thermometer bulb is submerged in the *water* (Figure 5.8). If it's not near 0° Celsius or 32° Fahrenheit, let it sit longer until it's close to that temperature. Leave the thermometer in the glass of ice water and check on it every five or ten minutes.

Figure 5.8

Does the temperature change while the ice is melting? Pay special attention to the point at which all of the ice melts. Does the temperature change *after* the ice melts? This whole process could take up to an hour, so be patient.

One last thing to do, and I highly recommend doing it because it's so cool! Get a candle, a crayon, and a business card. Use a pencil sharpener or a knife to make a bunch of crayon shavings and place them on top of the card. Light the candle and hold the business card right over the flame, as shown in Figure 5.9. Make sure you have plenty of shavings directly over the flame.

Hold the card there. After a while, the crayon will melt but the business card won't burn. Told you it was cool.

Figure 5.9

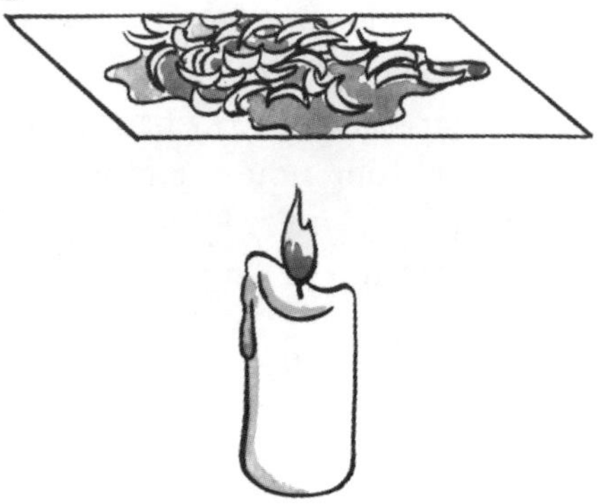

More science stuff

Let's start with what happened when you touched the objects just above the water line. The screwdriver and aluminum probably felt hot. The wooden spoon, though, didn't feel hot. The explanation for this is that the metals *conduct* heat very well and wood doesn't. It has to do with how atoms and molecules are put together in metals as opposed to wood.

There is a formula that tells you just how well a given material will conduct heat, so that given a temperature difference on either side of a piece of this material, you can calculate exactly how much heat will be conducted through the material in a given amount of time. I'm not going to bother you with that formula but I will tell you that each kind of material has a number associated with it called the **thermal conductivity** of the material. If that number is high, the material conducts heat very well and if the number is low, the material doesn't conduct heat well. Silver and copper have high thermal conductivities and rubber and wood have low thermal conductivities.

Air has a very low thermal conductivity, which is why you can insulate your house by using double- or even triple-pane windows. The layers of air in between the panes keep heat conduction, either in or out, to a minimum. Of course, if that's true, why do we use fiberglass insulation in the walls of a house? Why not just air? The reason is that, with most houses, any air that's in between the inner and outer walls isn't going to stay there very long. It readily escapes to the outside, to be replaced by colder or warmer air. The insulation in the walls of your house helps to *trap* as much air as possible, keeping it from moving around and transferring heat by convection.

Back to the objects in the water. When you pulled them out, dried them, and then touched them, the aluminum foil probably didn't even feel hot, while the screwdriver stayed hot for a long time. If you cook at all, you've probably experienced this before. If you take a heavy metal pot out of the oven, it takes a while to cool. You can touch an aluminum roasting pan, however, almost immediately after taking it out of the oven. But why should one kind of metal cool off faster than another? The key here is that these objects have different *masses*. The screwdriver has a lot more mass than the rolled-up aluminum foil. If you had a really thin sheet of the metal that the screwdriver is made of, it would cool off just as fast as the aluminum foil.

Both the screwdriver and the aluminum foil were about the same size, and when you took them out of the water, they were exposed to the same air. Stands to reason, then, that each lost heat to the surrounding air at about the same rate. If the objects lose heat at the same rate, then they will lose the same amount of heat in a given time period. So, the aluminum foil and the screwdriver lose about the same amount of heat in the time it takes to dry them off and touch them. Yet, the screwdriver clearly hasn't cooled off (changed its temperature) as much as the aluminum foil.[5] What this means is that *for a given amount of lost heat, objects with more mass change temperature less than objects with less mass.*

Now think about heating a pot of water. If you put an empty pot on the stove and heat it for one minute, it will get quite hot. But as we all know, if you put water in the pot and heat it for one minute, the water won't change its temperature much at all. You're adding the same amount of heat in both cases, but clearly the metal pot changes its temperature much more than the water does.

This means that different *substances* experience different changes in temperature for a given quantity of heat added or lost. There's a number that you can assign to different substances that expresses this fact and that number is called the **specific heat** of the substance. Substances with a high specific heat do not change temperature very much for a given quantity of heat added or lost. Substances with a low specific heat change temperature a lot for a given quantity of heat added or lost. Water has a high specific heat, so you have to add a lot of heat to change its temperature significantly. Metals have a low specific heat, so you can change their temperature rapidly by adding or taking away a small amount of heat.

So how much an object changes temperature for a given amount of heat added or lost depends on two things, the mass of the object and the specific heat of the substance that makes it up. There's a math relationship that expresses this and here it is:

(Heat added or lost) = (mass of object)(specific heat of object)(change in temperature)

[5] By the same token, the water transfers much more heat to the screwdriver than to the aluminum foil to get both of them hot in the first place.

Remember that parentheses next to each other mean multiplication. In symbol form, the relationship is:

$$Q = mc\Delta T$$

Q stands for the amount of heat added to or lost from the object, m is the object's mass, c is the specific heat of the material the object is made of, and ΔT is the change in temperature of the object (remember that Δ means "change in").

To see that this relationship explains our experiences, pretend you are adding a set amount of heat to an object. Then those three things on the right, multiplied together, have to equal that set amount of heat. What happens if you increase the mass of the object and keep its specific heat the same? Well, the quantity ΔT must get correspondingly smaller in order to still have the expression equal the same number.

Let's put in a few numbers to clear up those last few statements. Suppose you're adding 400 calories of heat to 200 grams of ice. When using these units for heat and mass, it's common knowledge that the specific heat of ice is 0.5.[6] To make sure the ice doesn't melt, we'll assume it starts at a temperature significantly below freezing. Then we have:

$$Q = mc\Delta T$$

$$400 = (200)(0.5)(\Delta T)$$

The value for ΔT that makes this a true statement is 4, meaning adding that amount of heat will change the temperature of the object 4° Celsius. Now what if we double the mass of the ice to 400 grams while keeping the heat added the same? Then we have:

$$400 = (400)(0.5)(\Delta T)$$

The value for ΔT that solves *this* equation is 2, meaning the more massive hunk of ice only changes its temperature 2 degrees for the same amount of added heat. Therefore, more massive objects change temperature less for a given amount of heat added.

Now suppose you replace the original 200 grams of ice with 200 grams of water. That means m stays the same and c gets larger. To be exact, the specific heat of water is 1.0.[7] With c getting larger, ΔT has to get correspondingly smaller.

---

[6] Of course this isn't common knowledge! Personally, I looked up the specific heat of ice in a reference book. You *could* figure the number out on your own with an elaborate experimental setup, but who wants to do that?

[7] The specific heat of water is, by definition, equal to 1.0. Like most sets of numbers assigned to physical quantities, specific heat requires an arbitrary starting point, and setting the specific heat of water equal to 1.0 is that starting point.

Let's use our number example to illustrate. The ice situation is the same as before.

$$Q = mc\Delta T$$

$$400 = (200)(0.5)(\Delta T)$$

As before, the change in temperature of the ice is 4 degrees. Now let's replace the ice with water. The specific heat increases to a value of 1. Now we have:

$$400 = (200)(1.0)(\Delta T)$$

The value of ΔT that solves this equation is 2, meaning the water, with a higher specific heat, changes its temperature only 2 degrees for the same amount of heat added. Therefore, objects with a greater specific heat change temperature less for a given amount of heat added.

So whenever you add heat to an object or take heat away from an object, it changes temperature, right? Wrong! When air from the room adds heat to a glass of ice water, the ice melts but the ice water doesn't change temperature until after the ice is melted. Then the water begins to change temperature. Hopefully you saw that when watching the temperature of the glass of ice water before and after the ice melted. Why doesn't the temperature change when the ice is melting? Because it takes energy to break apart the bonds between the molecules as the substance changes from ice to water. The added heat goes into breaking the bonds instead of increasing the kinetic energy of the molecules. If the average kinetic energy of the molecules doesn't change, then the temperature doesn't change.

It's not just ice that behaves this way. Any time any substance changes from a solid to a liquid, or from a liquid to a gas, any added heat goes into breaking apart molecules rather than changing the temperature. Of course, it also happens when a substance changes from a gas to a liquid or from a liquid to a solid. The only difference is that the heat is given off by the substance. For example, water *gives off* heat when changing to ice, and steam *gives off* heat when changing to water.

And now we're ready to understand the business card and the crayon shavings. Paper burns at a temperature of about 233° Celsius (451° Fahrenheit), but crayon melts at a much lower temperature than that. The candle transfers heat to the business card, but the business card transfers heat to the crayon shavings. When the crayon starts melting, its temperature no longer changes. All the heat added to the business card goes into melting the crayon, and nothing gets any hotter while this is going on. It never gets hot enough for the business card to burn.

Even more things to do before you read even more science stuff

Get a sheet of white construction paper, a sheet of black construction paper, and some tape. You'll also need a helper. Have your helper tape the two colors of construction paper to your hands, as shown in Figure 5.10.

Figure 5.10

Head back to your table lamp with the lampshade removed and ask your helper to turn on the lamp. Hold your hands an equal distance from the bulb, with the construction paper facing the bulb, as shown in Figure 5.11.

Figure 5.11

Wait a bit as your hands warm up from the heat coming from the bulb. Which hand feels hotter?

Even more science stuff

The activity with construction paper showed that some materials absorb and radiate heat better than others. Black construction paper absorbs radiated heat better and it also radiates heat better than white construction paper. The white construction paper reflects the heat that it doesn't absorb. That's why the hand with the black construction paper got hotter. Of course, this kind of thing doesn't happen only with construction paper. All kinds of different objects absorb and reradiate heat differently.

And now let's take a big conceptual leap and talk about heat coming to and heat leaving the Earth. The incoming heat comes from the Sun, and the reradiated heat heads out into space.

With all this heat coming into and leaving Earth, the Earth is clearly not a closed system. Even so, there must be some kind of balance between the incoming and outgoing heat. If not, the Earth would heat up or cool down rather rapidly.

Figure 5.12

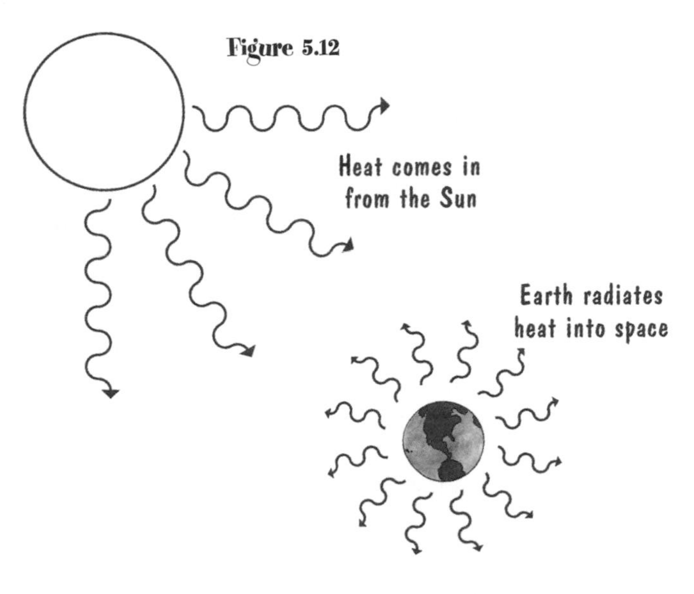

Figure 5.13

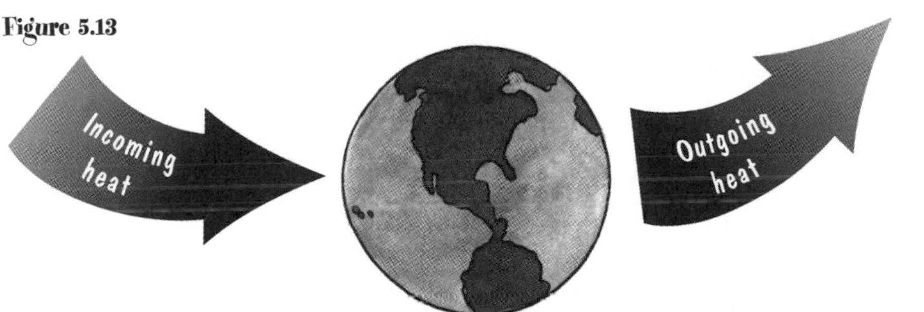

Hmmmm—Earth either heating up or cooling— sounds like global warming is rearing its ugly head. I'm not going to get deeply into the science of global warming and climate change. What I *will* do here is address some of the things that might affect the "heat balance" of the Earth, which is key to understanding the issue of global warming.

Clouds are one factor. Being white, they tend to reflect heat just as white construction paper does. So the more cloud cover there is, the less heat actually reaches the surface of the Earth. Various gases in the Earth's atmosphere also affect how much heat is reflected or passed on through.

Figure 5.14

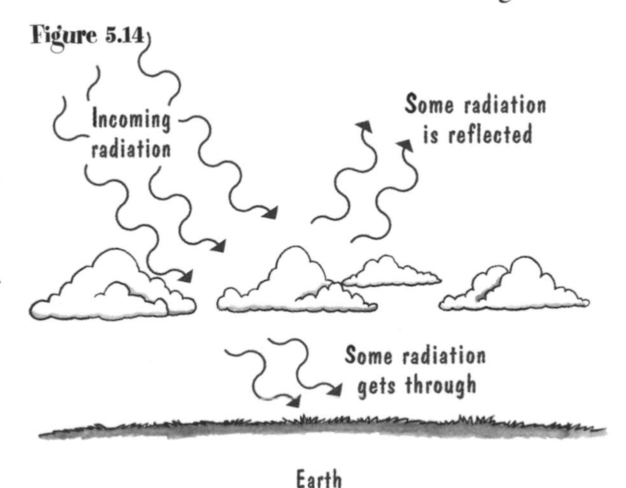

The Earth's surface is full of things that absorb heat and then reradiate it. The most abundant substance is water, which covers about three-fourths of the Earth's surface. Now the neat thing about water is that it has a high specific heat. And as you might remember, things with a high specific heat do not change temperature much for a given input or output of heat. All that water covering the Earth thus keeps the Earth from having wild fluctuations in temperature. If the oceans weren't around, we'd experience much hotter hots and much colder colds than we currently do.

Something else interesting happens when energy from the Sun interacts with things on the surface of the Earth. Radiated energy that comes from the Sun in a form that can pass through clouds and the atmosphere is absorbed by plants, dirt, and other objects on the Earth's surface. This energy is then reradiated in a slightly different form, which corresponds to infrared light or radiated heat. That heat *can't* pass through the clouds and atmosphere on the way out into space, so it's trapped here.

Figure 5.15

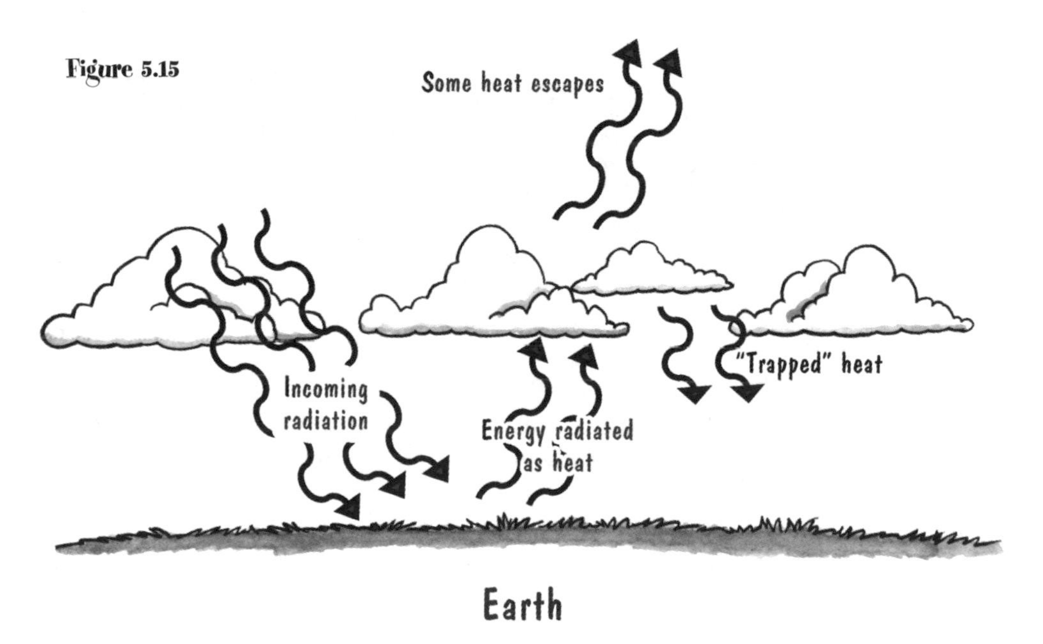

This is known as the **greenhouse effect**, which is kind of a funny name since it's not the main way that greenhouses stay warm! Greenhouses stay warm primarily by limiting *convection* of the air inside with the air outside. The incoming sunlight warms everything inside the greenhouse, including the air, and the walls of the greenhouse prevent the air inside from mixing with the cooler air outside.

Even though it's a misnomer, the greenhouse effect certainly does occur on the Earth, and that's a good thing. Without it, the surface of the Earth would be a whole lot colder than it is now. Obviously the kinds of gases in the Earth's atmosphere affect the warming that results from the greenhouse effect, and that's at the heart of debates over global warming. As you read about the subject of global warming, though, keep in mind that the greenhouse effect itself is not a bad thing. In fact, the greenhouse effect makes life, as we know it on Earth, possible. Too much greenhouse effect, though, and we would have a global climate more like that of Venus—not a good thing.

Chapter Summary

- Heat is a quantity of energy transferred from one system to another. Heat is usually measured in calories but it can also be measured in joules, just as any other kind of energy.
- The first law of thermodynamics describes the transfer of heat to and from a system. It's basically a statement of conservation of energy as applied to heat transfer.
- Heat can transfer from one system to another through conduction and radiation. Hot and cold liquids and gases can exchange places through the process of convection.
- Different substances have different thermal conductivities, which measure each substance's ability to conduct heat.
- Different substances also have different specific heats, which are a measure of how much the temperature of an object changes for a given quantity of heat added to or lost by an object. The mass of the object also affects the change in temperature for a given quantity of heat added or lost.
- Whenever a substance *changes state* (solid to liquid, liquid to gas, and the reverse of those), any heat added or lost goes into bringing about that change of state, with no corresponding temperature change.
- Different substances absorb and reradiate heat differently. This property affects the *heat balance* of the Earth and is a key concept for understanding the issue of global warming.

Applications

1. In a given room in your house, everything is pretty much the same temperature, right? Then why do some things feel cooler than others? In particular, metal objects feel cooler than fabric or wood, and smooth objects made from a given material feel cooler than rough objects made from the same material.

   To understand that, first realize that things at room temperature are cooler than your skin. That means that when you touch something, heat is going to transfer from your hand to the object. The faster heat transfers from your hand to the object, the cooler the object feels to you. Metal has a higher thermal conductivity than fabric or wood, so metal can conduct heat away from your hand faster than fabric or wood. Therefore, objects made of metal tend to feel cooler than objects made of fabric or wood.

   For a given material, smooth objects feel cooler than rough ones. That's because when you touch a smooth object, there are more points of contact between you and the object, resulting in a faster transfer of heat from your hand to the object. Whether it's due to material or smoothness or both, losing heat from your hand faster makes an object feel cooler.

Figure 5.16

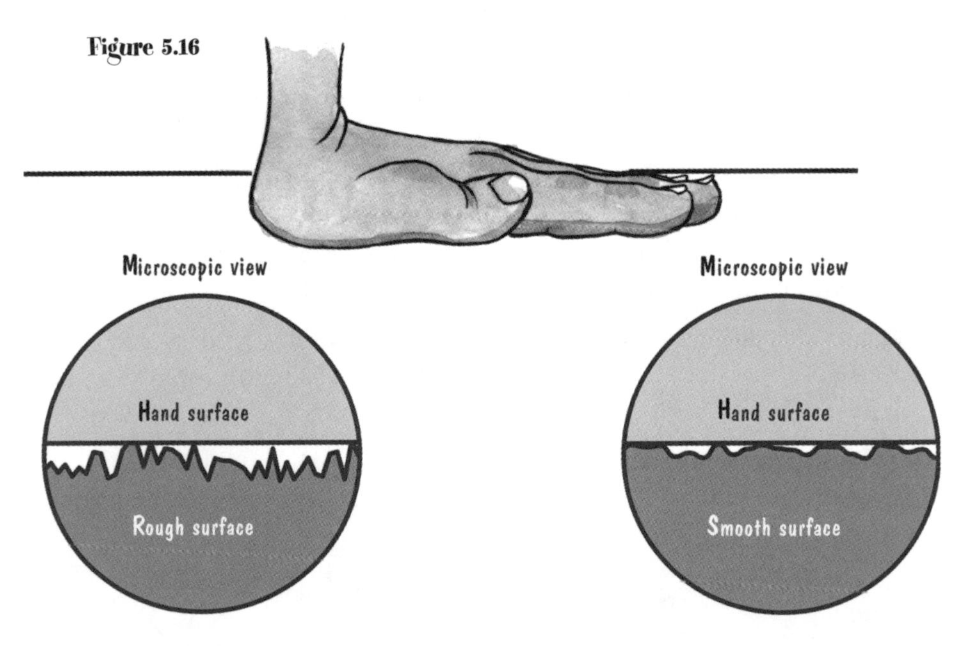

2. Heating systems usually have floor vents from which hot air flows. The entry vents are on the floor because hot air coming from them rises (pushed up by cooler air around it). Heating systems also have "return vents" that are

on the walls near the ceiling. What's the purpose of the return vents?

Return vents are built into heating systems so *convection cells* can be set up throughout the system. Cooler air returns to the furnace through the return vents as more warm air comes in through the floor vents.

3. One way to cool off on a hot day is to splash water on yourself. Why does that cool you off? Because as the water on your skin evaporates it requires energy (the water is changing from a liquid to a gas). The energy for that evaporation comes from your body, as heat flows from your body to the water. As you lose heat, your skin temperature goes down.

   You can feel even cooler if you splash rubbing alcohol on your skin because alcohol evaporates faster, drawing heat from your skin faster, and making you feel cool faster. That's why you give an alcohol bath to someone with a really high fever.

Figure 5.17

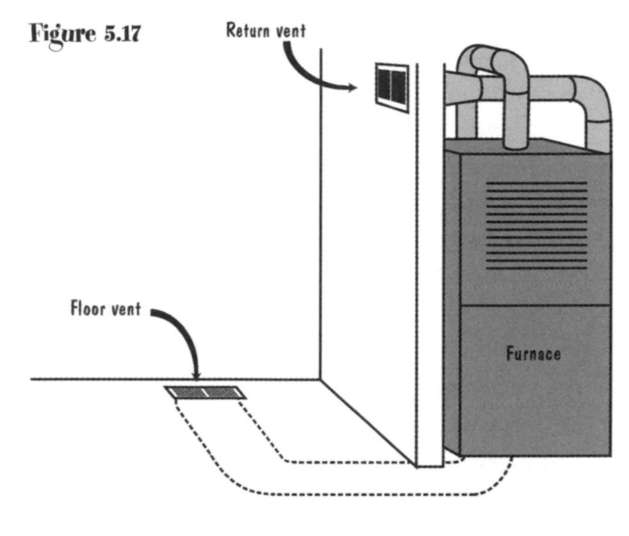

Figure 5.18

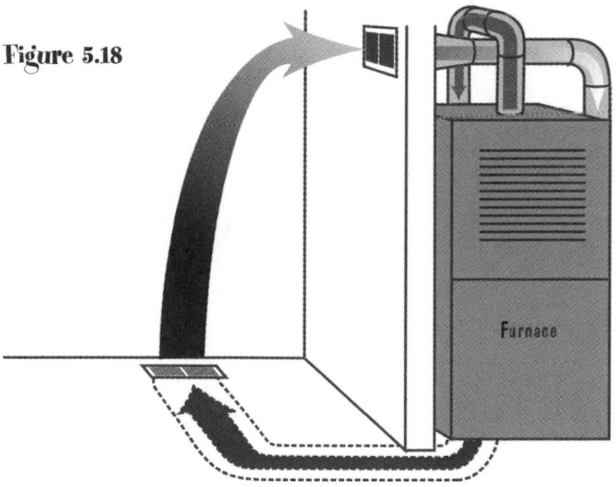

4. Ever wonder why climates near the ocean tend to be milder, with less dramatic temperature swings, than climates inland? Well, wonder no more. Because oceans have a high specific heat, they absorb heat from the Sun during the day without much change in temperature. At night, they release heat to the surrounding air without much change in temperature. Land, however, has a much lower specific heat than water, so it changes temperature much more for a given exchange of heat.

Taming Energy

Energy transformations take place all over the Earth without humans ever getting involved. Being the control freaks we are, though, we spend a lot of time trying to direct those energy transformations to make our lives easier. This chapter is about how we manipulate energy transformations to get electricity for important things like running model racing sets and model railroads.

Things to do before you read the science stuff

You've been sitting around reading long enough. Time to get some exercise. Find yourself a set of stairs and walk up slowly. Did your energy change? Answer: Yes. You gained (more precisely, you and the Earth gained) gravitational potential energy. How much gravitational potential energy did you gain? Answer: *mgh*, where *m* is your mass, *g* is 9.8 meters per second per second, and *h* is the vertical height of the stairs.

Figure 6.1

Go up the stairs a second time but this time run. Is it easier or harder to run up the stairs than to walk up the stairs? Don't tax your brain too long on this one—it's harder! Compare the gravitational potential energy gained when you run up the stairs with the gravitational potential energy gained when you walk up the stairs. Let's see, *m* is the same, *g* is the same, and *h* is the same. Therefore, the energy you gain is the same whether you run or walk. Why, then, is it more difficult when you run? Think about that for a while.

Because we're going to talk about electricity use, dig out a copy of your latest electric bill (don't look at the total—it will only upset you). If you have a solar-equipped house or live in an apartment with utilities paid, grab a bill from one of your friends. Find the place on the bill where it shows how many joules of electrical energy you used. Keep searching—it's there somewhere. Well, actually no, it's not. What you will find is the number of kilowatt-hours used (this is often abbreviated *kh* or *kwh*). Why is electrical energy measured in these strange units?

Now head outside and find the electricity meter for your house or apartment. There will be either a digital display (like the odometer on your car) or a series of wheels with numbers on them. Whatever kind of meter you have, notice that the units used to measure your energy use are, once again, kilowatt-hours.

The science stuff

You discovered in walking and running up the stairs that the *rate* at which you gain energy has some significance. If you do work on something, doing it quickly requires more effort than doing it slowly. To account for that, we define something known as **power**, which is the rate at which you do work or change energy.

$$\text{Power} = \frac{\text{work done}}{\text{time it takes to do that work}} = \frac{\text{change in energy}}{\text{time it takes for the energy change}}$$

Remember that when you do work on something, that amount of work is numerically equal to the change in energy of the object. That's why it's possible to substitute "change in energy" for "work done" in the above equation. In symbol form, this is:

$$P = \frac{W}{t} = \frac{\Delta E}{t}$$

where the symbol Δ still means "change in." Whatever the symbols we use, power represents the *rate* at which you do work. Because the work you do on an object equals the change in energy of the object, power is also equal to the change in energy divided by the time.

Let's apply this idea of power to you climbing a flight of stairs. To keep things simple, I'm going to assume you head up the stairs at a constant velocity. If your velocity doesn't change, then you don't change kinetic energy. Your total change in energy is just your change in gravitational potential energy. That's equal to *mgh*, where *m* is your mass, *g* is equal to 9.8 meters per second squared, and *h* is the vertical height of the stairs. Whether you climb the stairs in a short time or a long time, your change in energy is the same and is equal to *mgh*.

The power involved, however, is different for different times. I'll use a few numbers to show that. Let's suppose *mgh* in this example equals 1500 joules, which corresponds to a 75 kilogram (165 pound) person climbing stairs with a vertical height of 2 meters. If you climb these stairs in 5 seconds, the power involved is:

$$P = \frac{\Delta E}{t} = \frac{mgh}{t} = \frac{1500 \text{ joules}}{5 \text{ seconds}}$$

This equals the number 300 (I'll talk about the units of power in just a bit). Now suppose you climb those same stairs in 10 seconds. Then we have:

$$P = \frac{\Delta E}{t} = \frac{mgh}{t} = \frac{1500 \text{ joules}}{10 \text{ seconds}}$$

which equals the number 150. We now have half the amount of power generated. In other words, the faster you change your energy, the more power you generate.

Now let's look at the same situation from a "work" perspective rather than an energy perspective. Exactly who or what is doing the work on whom or what? Believe it or not, the stairs are exerting a force on you.[1] As the stairs exert a force on you, they do work on you. To calculate the amount of work the stairs do on

---

[1] This has to do with something called Newton's third law. When you push on the stairs, the stairs push back on you. For a detailed discussion of this, see the *Force and Motion* book in this series.

you, we use our trusty formula for work, which is $W = Fd_{||}$. When you use this formula, you find that the work done in moving you up the stairs at a constant velocity is just equal to *mgh,* where *h* is the height of the stairs, *m* is your mass, and *g* is 9.8 meters per second squared.[2] This quantity is the same whether you move slowly or quickly, just as the change in energy is the same whether you move slowly or quickly. The power generated is work divided by the time to do that work, or:

$$P = \frac{W}{t}$$

If *t* is large (moving slowly), the power generated is small. If *t* is small (moving quickly), the power generated is large.[3] The faster you do work, the more power you generate.

The unit of power is the *watt,* named after James Watt; 1000 watts gives you one kilowatt. Okay, we're getting closer to understanding what a kilowatt-hour is. You measure the rate at which you use electrical energy in kilowatts because that's a unit of power. But you don't pay your electrical bill based on the rate you use the energy; you pay on the total amount of energy used.

Before explaining how kilowatt-hours are units of energy, I'm going to take you on a diversion. Suppose you want to measure how many miles you've traveled in a car, knowing how fast you've traveled and for how long . What you would do is multiply miles per hour by the number of hours driven.

$$\text{Distance traveled} = \frac{\text{miles}}{\text{hour}} \times \text{hours}$$

Once you get your result, you could say that you have measured the distance not in miles but in (miles/hour)-hours. Of course that would be a totally silly thing to do, but you could do it.

Similarly, to calculate the total energy used when you know the rate at which it is being used (which is power, which is the energy used divided by the time), you multiply the rate (energy divided by time) by the number of hours you used the energy at that rate.

$$\text{Energy used} = \frac{\text{energy used}}{\text{time}} \times \text{time}$$
$$= \text{power} \times \text{time}$$

[2] Refer back to Chapter 2, where I show that this is the work done when lifting an object through a height *h*.

[3] If that's not obvious to you, just put in some numbers for *W* and *t*, just as I did in the previous paragraph with ΔE and *t*.

So a kilowatt-hour is just a unit of energy. Why, exactly, power companies use kilowatt-hours instead of joules, I don't know. I'm betting that, like many other things, it's just a tradition that made sense a long time ago and has just hung around. Or maybe energy companies used to concern themselves with the rate at which people used energy, and then an accountant came along and told them to hang the rate of use and just worry about what was really costing them—the total amount of energy used.

More things to do before you read more science stuff

What I'm going to have you do in this section is a bit more complicated than what I usually request. It's really instructive, though, so you just might find the extra effort worthwhile. Anyway, you have to start by gathering a few materials. You need the following:

- A compass (the kind that tells you what direction north is). The larger the compass the better. The really tiny ones (a centimeter or two in diameter) won't work.
- A strong magnet (the stronger the better—refrigerator magnets won't cut it this time).
- A 2-meter (approximate) section of *insulated* wire. This needs to be thin enough so that you can bend it easily, yet strong enough that it will hold its shape. The folks at the hardware store will be able to help.
- A working flashlight battery.
- Some kind of tape.
- A tube from a roll of toilet paper, sans toilet paper.

Start by stripping a small bit of insulation off each end of the wire. Then wrap the wire a bunch of times around the compass. Leave about 10 centimeters of wire on one end and at least a meter of wire on the other end. After making sure you can still see the compass needle underneath all that wire, tape the wire in place. See Figure 6.2.

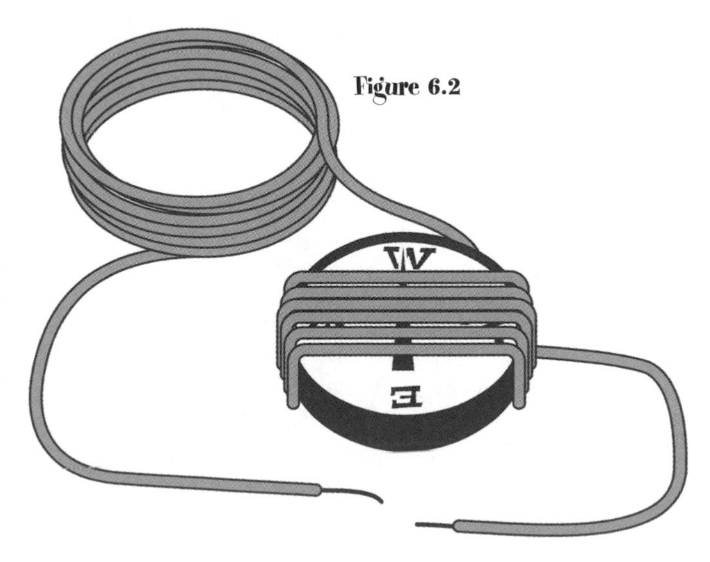

Figure 6.2

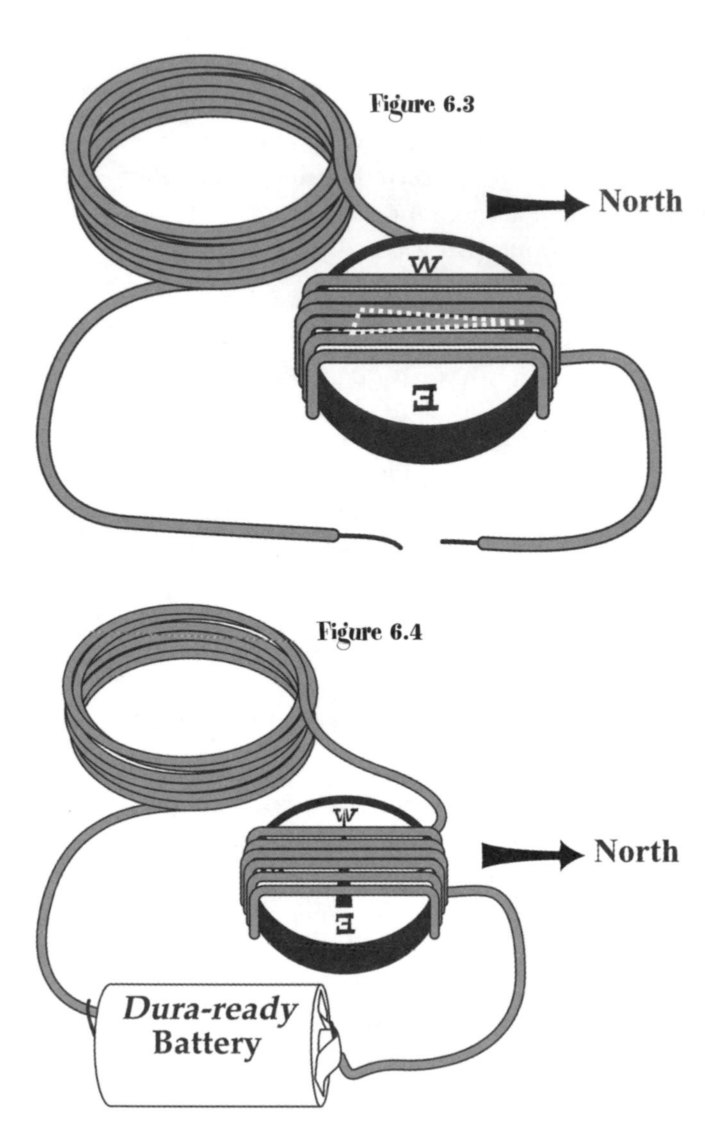

Figure 6.3

Figure 6.4

Make sure the compass isn't near your magnet, or any other magnet for that matter (phones and stereo speakers have pretty strong magnets in them). Place the wrapped compass on a flat surface. The compass needle should point north. Rotate the compass until the needle is lined up with the coil of wire (Figure 6.3).

With the compass in that position, touch the free, stripped ends of the wire to the battery terminals, as shown in Figure 6.4. Watch what happens to the compass needle when you do this.

If the compass needle doesn't deflect, check to see that the battery isn't dead and that you have the compass lined up as in Figure 6.3. With a good battery and the correct alignment, the compass needle will deflect as shown in Figure 6.4. Now batteries cause electrical current to flow in wires, so obviously having an electrical current going through the coil on the compass causes the needle to deflect. Don't worry about why that happens, because it would take half a book to explain it well!

Set the battery aside and cut the toilet paper tube in half. Wrap the long free end of the wire around one of the half tubes as shown in Figure 6.6. Leave enough wire so you can twist together the stripped ends and be able to place the tube at least 30 centimeters from the compass.

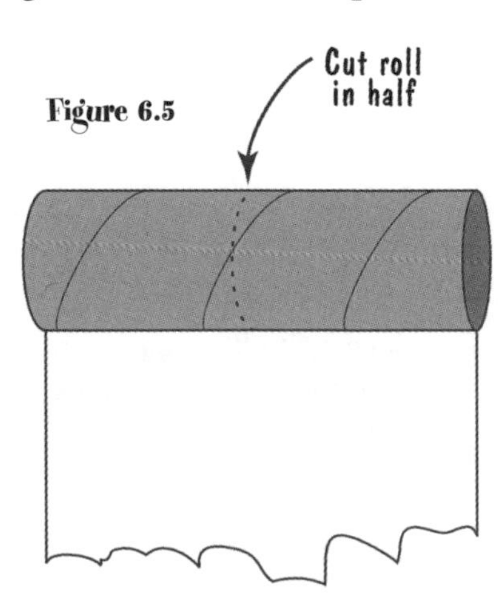

Figure 6.5

Figure 6.6

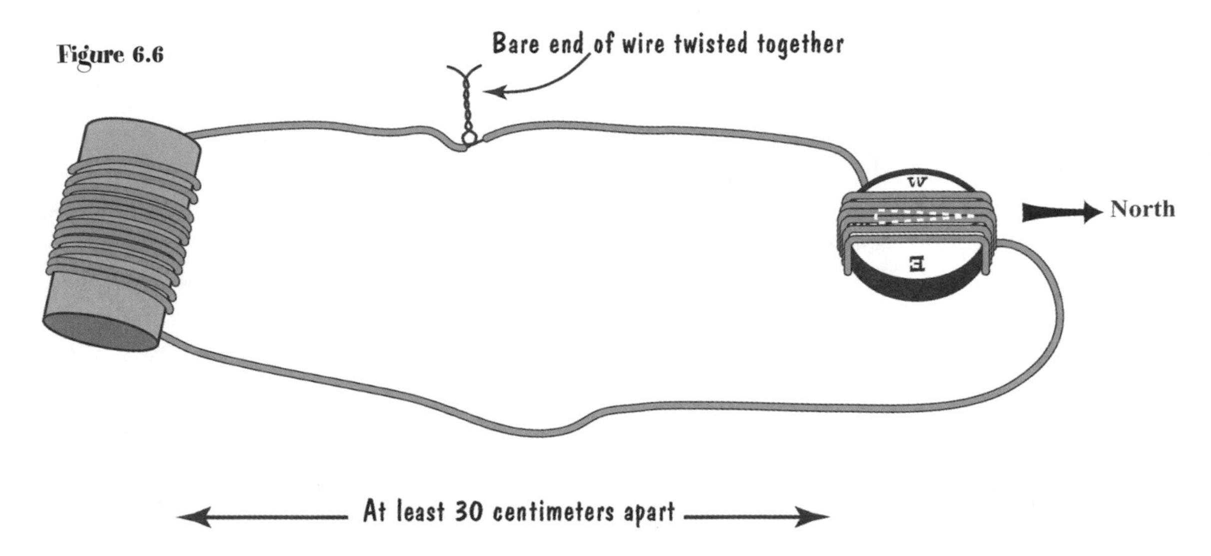

Move the magnet in and out of the center of the tube. As you do this, watch the compass needle. (See Figure 6.7.)

Figure 6.7

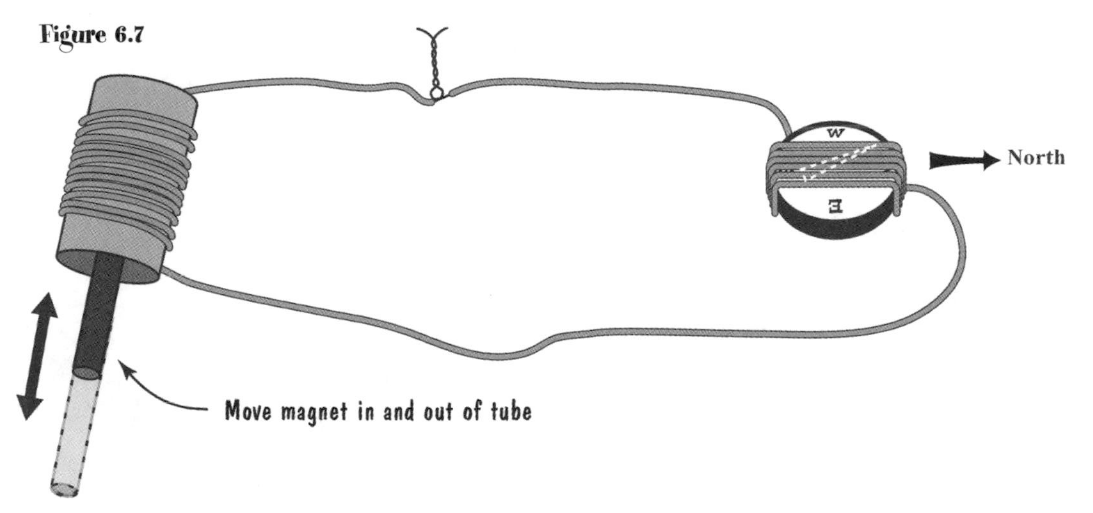

The compass needle should move just a tiny bit and only while the magnet is moving. Don't expect the needle to move as much as it did when you touched the wires to the battery.

More science stuff

See if the following reasoning makes sense. The compass needle deflects whenever an electrical current is flowing through the coil of wire that's wrapped around the compass. The needle deflects when you move the magnet in and out of the coil of wire wrapped around the toilet paper tube. Therefore, moving the magnet in and out of the coil of wire causes an electrical current to flow in the wire.

If that reasoning doesn't make sense, well, just pretend that it does. It's a fact of nature that, when you move a magnet in and out of a coil of wire, you will generate an electrical current in the wire. Also, if you move a coil of wire relative to a magnet, you will also generate an electrical current. It turns out that this is the method used to generate just about all the electricity people use. Before going on, I'll mention the other two main sources. One is the battery, which takes chemical potential energy and converts it to electrical energy. The other is the solar cell, which takes radiant energy from the Sun and converts it to electrical energy.

Figure 6.8

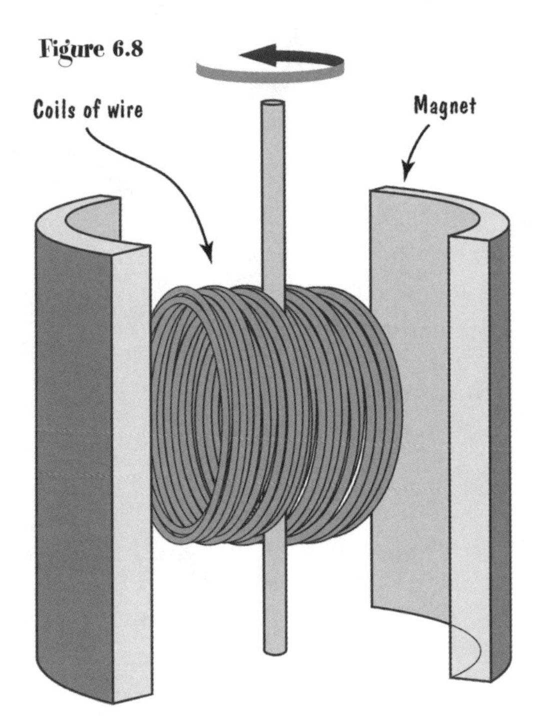

The device used to generate electricity is called a **turbine**. A turbine basically consists of many coils of wire that are free to spin inside stationary magnets (Figure 6.8).

As the coils spin, they move relative to the magnets, resulting in an electrical current flowing through the wire. That electrical current then can be sent to consumers. More about that later.

So all we need to do is hire a bunch of people at minimum wage to spin turbines, right? Well, no, you couldn't hire enough people to meet the demand for electricity. Better to get an input of energy from some other source. One source is a moving river. Set your turbines up on a dam in a river and let the moving water turn the turbines. What you get is **hydroelectric power** (Figure 6.9).

Figure 6.9

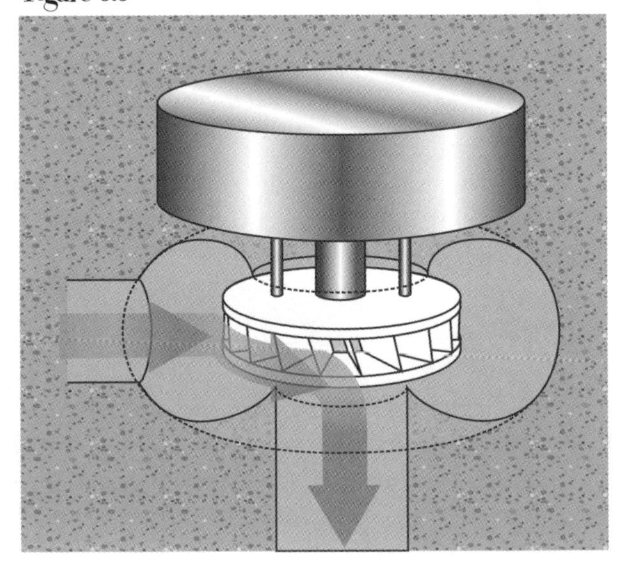

Another source of energy for turning turbines is the wind. Set up a windmill connected to a turbine, and the wind will turn the turbine for you (Figure 6.10).

Figure 6.10

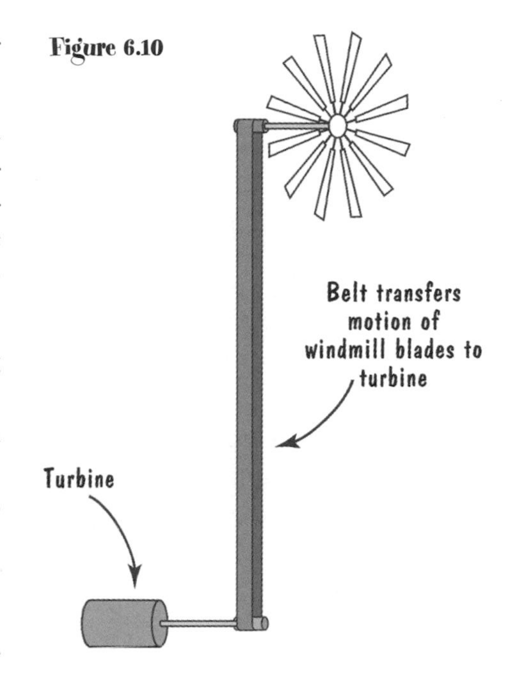

One trouble with hydroelectric and wind power is that they can be unreliable. No wind—no electricity. Little rain means lower water levels and less hydroelectric power. To have a more reliable source of electricity, we use the setup shown in Figure 6.11.

You heat the water reservoir, producing steam. You then force that steam through pipes that lead to—surprise—a turbine! The forced steam turns the turbine, producing electricity. Simple enough. Now all you need is a way to heat the water. There are lots of ways to do that. Burning coal is one way. Burning natural gas is another. Using the heat from nuclear reactions is another. This last method is somewhat controversial, in part because people don't understand what role the nuclear reactions in a nuclear reactor play in generating electricity.[4] Their sole purpose, though, is to generate enough heat to turn the water to steam, which turns the turbine, which generates the electricity, which goes to the old lady who swallowed the fly.

Figure 6.11

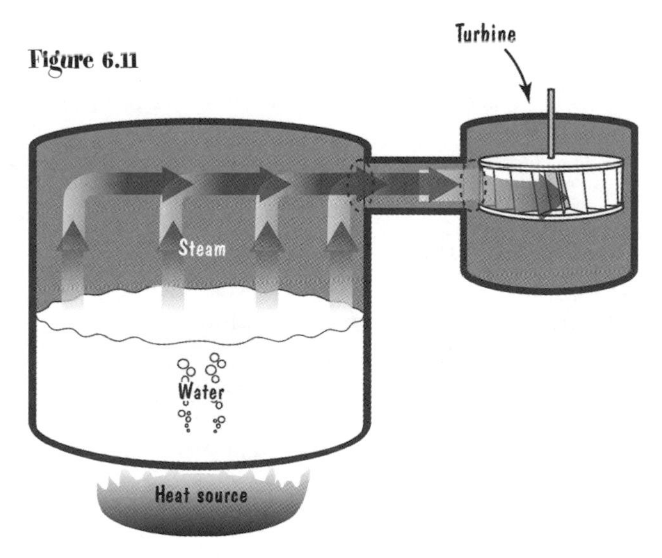

As a summary, let's try to draw a diagram that shows all the energy transformations that result when burning coal to get electricity to the consumer. In the diagram, I'm going to put in "losses" to thermal energy in lots of places.

One place you might not expect such losses is in getting electricity from the power plant to the consumer. Any time electrical current flows in wires, however, it generates heat. The

[4] Take note of the words "in part." I'm not saying this is the only reason. I just don't want to get into all the other issues, such as waste disposal, here.

scientific explanation for what's going on is that very tiny electrons bang into atoms and produce heat. The details of that scientific model will have to wait for the *Stop Faking It!* book on electricity. Anyway, here's the diagram.

Figure 6.12

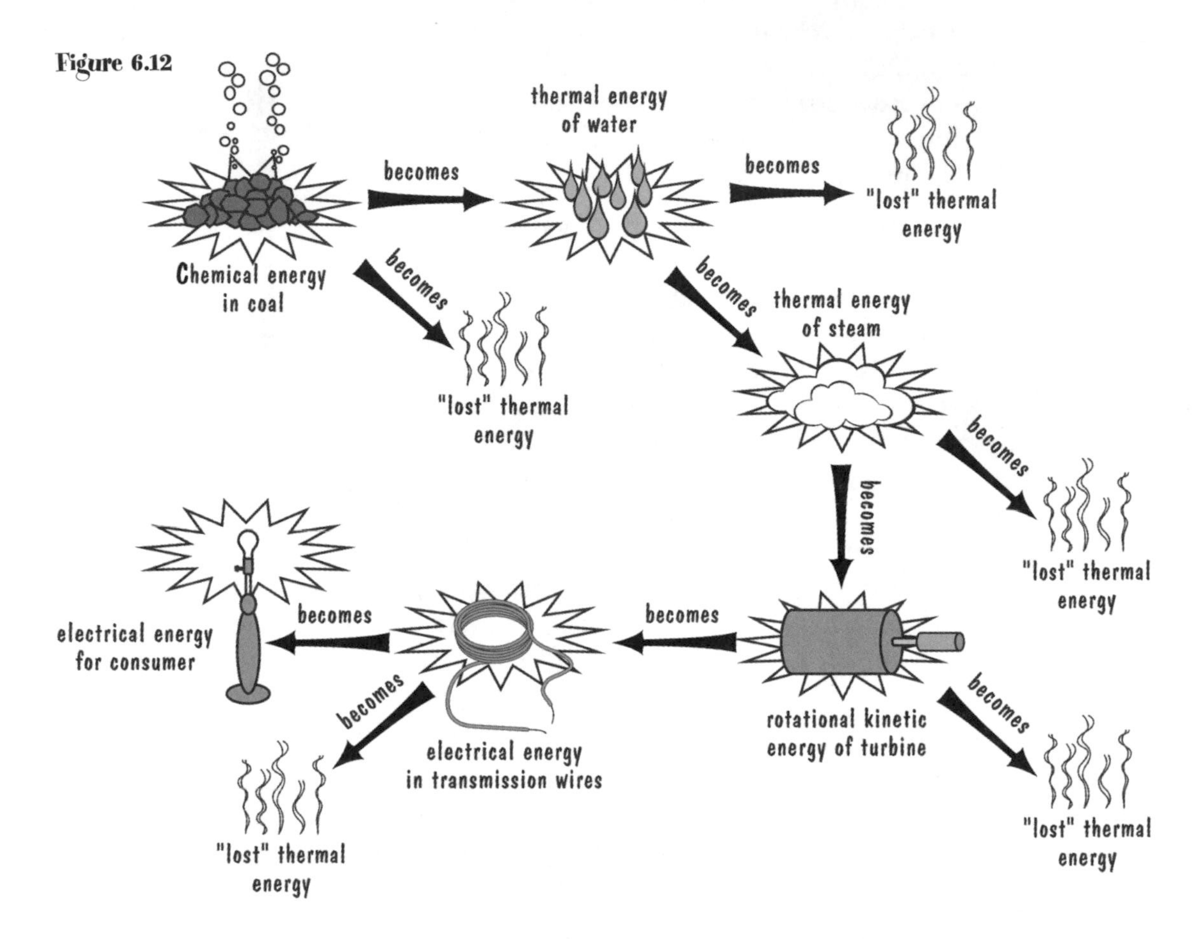

Well, that's about it for energy basics. We certainly haven't covered all the applications, because energy and energy transformations are at the heart of many different areas of physics and many different scientific disciplines. To get a full appreciation of how energy concepts apply in these other situations, take a look at the other books in the *Stop Faking It!* series. How's that for ending a book with an advertisement!

Chapter Summary

- Power is the rate at which you do work or use energy. Power is measured in watts.
- Kilowatt-hours is a unit of energy, not power, and is how the power company measures energy usage by consumers.
- When wires and magnets move relative to each other, an electrical current flows in the wires. This phenomenon is the basis for most of the electricity generated in our country. We use various methods to turn a turbine (coils of wire turning inside a bank of magnets) to generate electricity.

Applications

1. What makes electricity cost more in one state than in another, or even in different parts of the same state? Many factors affect that, including the number of nearby power plants, the regulations on those power plants, and just a whole bunch of politics. I won't deal with those, but I will discuss the factor of *efficiency*. If it's possible to get electricity to the consumer with as few energy "losses" as possible, then the electricity should be cheaper. Those energy losses could come at the power plant, where inefficient equipment (machines!) results in more energy losses. Also, the farther you have to transmit the electricity, the more you lose heat as the electricity travels through more wires. That means consumers in remote locations generally pay more for their electricity than urban consumers.
2. Speaking of efficiency, how come gas appliances are so much more efficient (cheaper to operate) than electrical appliances? Consider a clothes dryer. With a gas dryer, you get natural gas piped into your home and burn it to produce the heat to dry your clothes. With an electric dryer, the power company burns natural gas or coal to produce electricity, which you then use to create heat to dry the clothes. It's a whole lot cheaper to use the raw materials (gas or coal) directly rather than have it converted to electricity first. Of course, it's even cheaper to use the ultimate source of energy that created the natural gas and coal. That source is the Sun, but check your neighborhood for ordinances before hanging your clothes outside to dry!

SCiLINKS®
THE WORLD'S A CLICK AWAY

Topic: alternative forms of energy

Go to: *www.sciLINKS.org*

Code: SFE17

Glossary

bimetallic strip. A strip made of two different kinds of metal. Because different metals expand and contract different amounts when their temperature changes, a bimetallic strip bends and straightens with changes in temperature. Bimetallic strips are an important component of dial thermometers and thermostats.

block and tackle. An arrangement of fixed and moving pulleys, combined with rope or wire, which enables one to lift very heavy objects with a small input force.

conduction. The transfer of heat as a result of direct contact between the molecules of different substances.

convection. The transfer of heat through the mixing of molecules of different temperatures. This transfer often takes place in a circular path known as a convection cell.

conservation of energy. A principle stating that in a closed system, the total amount of energy remains constant.

efficiency. The work output of a machine, divided by the work input to the machine, times 100%. This is a measure of how much energy is "wasted" in the operation of a machine. Also something about which government agencies know very little.

energy. An abstract concept that is difficult to define concisely. One definition is that energy is "the ability to do work." Another is "what an object gains when work is done on it." Still a third is that energy is "the quantity that is conserved in the principle of conservation of energy."

first law of thermodynamics. A statement of conservation of energy as it applies to the process of adding heat to or removing heat from a system. In symbolic form, it is $Q = \Delta U + W$, where Q is the heat added to a system, ΔU is the change in internal energy of the system, and W is the work done by the system.

force. A push, pull, hit, nudge, bump, or whack.

fulcrum. The point about which a lever rotates. Most simple levers or combinations of levers, such as pliers, scissors, and bottle openers, have a fixed fulcrum. Others, such as a crowbar, have different fulcrum positions for different applications.

gravitational potential energy. The potential energy contained in the combined system of an object and the Earth by virtue of the separation between the object and the Earth's surface. The expression for gravitational potential energy when the object is near the Earth's surface is mgh, where m is the mass of the object, g is a number equal to 9.8 meters per second per second, and h is the distance the object is above some arbitrary reference level. For objects not near the Earth's surface, there is a more general expression for gravitational potential energy—one that involves the separation between the object and the center of the Earth.

greenhouse effect. A name given to the process in which radiation enters through a barrier (e.g., glass or clouds) and is then reradiated in a form that cannot pass back through the original barrier. This process results in the trapping of heat radiation and the warming of the area inside the barrier. Interestingly enough, greenhouses do not rely on the greenhouse effect as a primary means of staying warm in cold weather.

heat. A quantity of energy transferred from one system to another because of temperature differences. Heat gained or lost by a system tends to change the internal energy of the system.

internal energy. The total of all the kinetic and potential energy contained in a system. This is also known as thermal energy.

joule. The unit in which energy is measured in the Système Internationale (SI) system of units.

kinetic energy. The energy something has because it is in motion. This energy is often separated into translational (linear) kinetic energy and rotational kinetic energy. The kinetic energy of something moving but not rotating is given by $\frac{1}{2}mv^2$, where m is the mass of the object and v is the velocity of the object.

kinetic theory of gases. A scientific model for how the molecules in a gas move and interact with one another. The model states that the motion of the molecules is random, and that no energy is "lost" when the molecules collide with one another and with their surroundings.

lever. A simple machine that consists of a straight piece of solid material that's free to rotate about a point called the fulcrum. Depending on where the fulcrum is, a lever can output a larger or a smaller force than the force you apply to it.

perpetual motion machine. A machine that operates continuously without any input of energy. Such a machine would violate conservation of energy. Not surprisingly, no one has ever successfully built one of these. No doubt you can find many people claiming to have built a perpetual motion machine with one search on the Internet.

potential energy. The energy an object or group of objects has by virtue of the object's shape or the relative positions of the objects in the group.

power. The rate at which work is done or at which energy is gained or lost. It's defined as work divided by time or energy divided by time.

pulley. A grooved wheel on an axle. A rope, string, or wire fits in the groove so that when the rope moves, the wheel turns. When using more than one pulley in a combination, you can transform a small input force into a large output force.

radiation. The transfer of heat (and actually many other kinds of energy, such as light and X rays) by the transmission of electromagnetic waves. Unlike other processes of heat transfer, radiation can travel through empty space.

reference level. An arbitrarily chosen position from which the gravitational potential energies in a given situation are measured.

rotational kinetic energy. The kinetic energy an object or group of objects has by virtue of the fact that they are rotating. The formula you will see in physics textbooks for rotational kinetic energy is $\frac{1}{2}I\omega^2$, where I is something known as the *moment of inertia* and w is known as angular velocity.

specific heat. A number associated with each substance that tells how much the substance's temperature changes for a given quantity of heat added or taken away.

temperature. How hot or cold something is. The temperature of an object is related to the average kinetic energy of the molecules in the object.

thermal conductivity. A number associated with each substance that tells how easily that substance conducts heat.

thermal energy. Another name for internal energy. Head over to that definition, cut it out, and paste it here.

translational kinetic energy. The kinetic energy an object has by virtue of its motion in a straight line. This concept is used only when you decide to separate an object's kinetic energy into translational and rotational kinetic energy.

turbine. A contraption that, when rotated by some external force, generates electricity. The key components in turbines are coils of wire and magnets.

watt. The unit of power, equal to joules/second, in the SI system of units. Also, a question commonly heard in science classes.

work. The force applied to an object, multiplied by the distance the object moves in the direction of that force. This scientific definition has little to do with what we normally think of as work. It is said that when people are on their deathbeds, they seldom say they wished they had spent more time at work instead of with their families, hobbies, or recreation. Those people are clearly not talking about the scientific definition of work.

Index

Page numbers in **boldface** indicate figures.

E

J

K

L

M

N

P